D1415562

Lecture Notes in Earth Sciences

Edited by Somdev Bhattacharji, Gerald M. Friedman, Horst J. Neugebauer and Adolf Seilacher

9

Gisela Gerdes
Wolfgang E. Krumbein

Biolaminated Deposits

Springer-Verlag

Berlin Heidelberg New York London Paris Tokyo

Authors

Dr. Gisela Gerdes
Prof. Dr. Wolfgang E. Krumbein
Geomicrobiology Division, University of Oldenburg
Carl-von-Ossietzkystr. 9–11
D-2900 Oldenburg, West Germany

ISBN 3-540-17937-2 Springer-Verlag Berlin Heidelberg New York
ISBN 0-387-17937-2 Springer-Verlag New York Berlin Heidelberg

Printed in Germany

Printing and binding: Druckhaus Beltz, Hemsbach/Bergstr.
2132/3140-543210

Petrificata montium calcariorum non filii sed parentes sunt, cum omnis calx oriatur ab animalibus (Linnaeus, Systema Naturae, Ed. XII, T. III, p. 154, 1760-1761)

PREFACE

The geological significance of life has long attracted mankind. Not only have single groups of organisms been considered, such as frame-building animals, diatoms or "monera" (radiolarian, globigerins, forams), but unitarian pictures were also drawn concerned with the regulation and feedback of geochemical cycles by interacting metabolic pathways. The enzyme-controlled back-coupling system of living and inanimate matter fascinated Vernadsky (1863 - 1945), a mineralogist and crystallographer, and is again stressed in Lovelock's Gaia hypothesis and Krumbein's Bioplanet or Bioid approach. The role of microorganisms in this respect is well documented in terms of disintegration of rocks, production and mineralization of organic compounds, catalyzation of the oxidation and reduction of metals, biomineral formation and biogenic ore formation. Records of stromatolites arising from the vital activity of microorganisms date back to the earliest known sedimentary environments of the Precambrian era.

The aim of the work presented here is to document the in-situ stratified accretion of sediments attributable to the vital activity of microbes. Part I comments on terms which relate to microbially produced sedimentary structures and products. Part II is concerned with the occurrence of microbial mats (potential stromatolites) in modern marginal marine environments of arid and temperate coastlines. Varying modes of facies evolution in subenvironments are shown through the integration of sedimentological, microbiological and faunistic data. In Part III structures attributed to the activity of Precambrian, Permian and Lower Jurassic microbial communities are analyzed, and some complementary aspects concerned with the geological potential of microbes are summarized.

Acknowledgements (Gisela Gerdes)

Presented here is a modified version of my thesis which encompasses a number of individual publications. I am indebted to many people who accompanied my way over the past years. My benefactor in this work was W. E. Krumbein. He first introduced me to the fascinating system of microbial mats. From Gavish Sabkha and Solar Lake we went on to include the "Farbstreifen-Sandwatt" as parts of the expanding biosedimentary system. We then turned our attention to counterparts of all this in fossil records, spanning the gap between biology and geology.

My first encounter with actuopaleontology was during my cooperation with W. Schäfer. His book "Aktuopaläontologie nach Studien in der Nordsee" was the first scientific work which I was able to follow through from its conception. His "Schule des Sehens", which was transformed into reality through the reorganization of exhibits at the Senckenberg Museum, Frankfurt, remains one of the most memorable impressions of my stay in that city.

H.-E. Reineck provided support and advice in the fields of actuogeology and actuopaleontology. Our collaboration began in "Senckenberg am Meer", Wilhelmshaven. I would like to thank him for the interest he shared in my work and for all his help and advice. During our trips to ancient and modern depositional environments and through our work in the laboratory he taught me to recognize and understand sedimentary structures.

My thanks are further extended to my other benefactor, H. K. Schminke. I am grateful also to colleagues from the Geomicrobiology team and to K. Wonneberger, my former partner at Oldenburg University marine biology unit, Wilhelmshaven, for their discussion and advice. Memories of our work together on Mellum, in the Gavish Sabkha, by Solar Lake and in Elat unite me with E. Holtkamp. Our stay, laboratory work and accommodation on Mellum were made possible by the Mellum Council and in Israel by the H. Steinitz Marine Biology Laboratory, Elat and its staff. I am particularly grateful to F. D. Por for his advice during our stay in Israel. I would also like to thank all for assistance and care in the preparation of drawings, reproductions, photographs, thin sections and checking of the manuscripts: R. Flügel, G., K. Oetken and H. Gerdes, W. Golletz, A. Grünert, E. Johnston, M. and H. Müller, I. Raether, V. Schostak, L. Tränkle. I especially want to thank J. Gifford for her patient help in transforming this manuscript into readable English.

Finally, I am indebted to Dr. Engel and Springer Verlag for publication in the Lecture Notes series. I would like to thank everybody who made this possible.

SUMMARY

Biolaminated deposits, produced by microbial communities, were studied in modern peritidal environments and in the rock record. The term microbial mat refers to modern, the term stromatolite to ancient analogs. The term biolaminated deposits was used to encompass both microbial mats and stromatolites.

Microbial mat environments studied are the Gavish Sabkha, the Solar Lake, both hypersaline back-barrier systems at the Gulf of Aqaba, Sinai Peninsula, and the "Farbstreifen-Sandwatt" (versicolored sandy tidal flats) on Mellum, an island in the estuary embayment of the southern North Sea coast. Three facies-relevant categories were distinguished: (1) the mat-forming microbiota, (2) environmental conditions controlling mat types and lithology, (3) bioturbation and grazing.

Cyanobacteria account for biogenic sediment accretion in all cases studied. Three major groups occur: filamentous cyanobacteria, coccoid unicells with binary fission and those with multiple fission. In the presence of these groups the following mat types evolve: (1) continuously flat (stratiform) L_h-laminae (occur in all environments studied); (2) translucent, vertically extended L_v-laminae (only Gavish Sabkha and Solar Lake); (3) nodular granules (only Gavish Sabkha).

Basically, the development of mats is controlled by moisture. Thus high-lying parts where the groundwater table runs more than 40 cm below surface are bare of mats. These are: The circular slope and elevated center of the Gavish Sabkha, the shorelines of the Solar Lake and the episodically flooded upper supratidal zone of Mellum Island. The following situations of water supply were found to stimulate mat growth: (1) Capillary movement of groundwater to exposed surfaces, (2) shallowest calm water, both realized in the Gavish Sabkha and the Solar Lake. On Mellum Island, mats form in the lower supratidal zone, which is flooded in the spring tide cycle and wetted during low tide by capillary groundwater. Salinity is almost that of normal seawater, whereas in the Solar Lake, it ranges from 45 $^{o}/oo$ to 180 $^{o}/oo$ and in the Gavish Sabkha, it reaches more than 300 $^{o}/oo$. Salinity increase is correlated with rising concentrations of magnesium and sulfate ions.

In the Gavish Sabkha, episodic sheetfloods cause high-rate sedimentation which is accidental to the living mats. Episodic low-rate sedimentation stimulates the mats to grow through the freshly deposited sediment layer. This occurs predominantly on Mellum Island due to eolian transport.

Within the Gavish Sabkha, mineralogy of sediments, community structures, standing crops, redox potentials and pH are highly correlative to the increasing evenness in moisture supply which is realized by the inclination of the system below mean sea level. These conditions bring about a lateral sequence of facies types which include (1) siliciclastic biolaminites at the coastal bar base, (2) nodular to biolaminoid carbonates at saline mud flats, (3) regularly stratified stromatolitic carbonates with ooids and oncoids within the hypersaline lagoon, (4) biolaminated sulfate toward the elevated center. High-magnesium calcite in facies type 3 precipitates around decaying organic matter and forms also the ooids and oncoids. These occur predominantly within hydroplastic L_v-laminae which provide numerous nucleation centers.

Within the Solar Lake, facies type 3 (stromatolitic carbonates with ooids and oncoids) is most important, and grows to extraordinary thickness at the lake's shelf. The regular alternation of dark and light laminae results from seasonally oscillating water depths. These conditions couple back over changing light and salinity intensities to changing dominance structures of mat-building communities. Increasing salinity correlates with decreasing water depth and accounts for the relative abundance of coccoid unicells and diatoms, both active producers of extracellular slimes (L_v-laminae). Water depths locally or temporarily increased favor surface colonization by *Microcoleus chthonoplastes* (L_h-laminae).

The biolaminated deposits of the versicolored tidal flats on Mellum Island are similar to facies type 1 of the Gavish Sabkha (siliciclastic biolaminites). Differences exist in the lithology: Sediments upon or through which the mats on Mellum Island grow are made up of clean sand. The grains originate predominantly from re-worked glacial sediments and are rounded to well rounded. By contrast, the strong angularity of siliciclastic grains in the Gavish Sabkha clearly shows their status as primary weathering products.

In all environments studied, insects play a significant role. Mainly salt beetles contribute to the lebensspuren spectrum. There is no indication that burrowing and grazing beetles and dipterans are detrimental to the growing mat systems. According to the marine fauna, two distributional barriers exist: (1) physical and (2) biogeochemical factors. Physical barriers are (a) hypersalinity and barrier-closing, which restrict the marine fauna in the Gavish Sabkha and the Solar Lake to a few species, mainly meiofaunal elements such as ostracods and copepods. Only in the Gavish Sabkha, one marine gastropod species occurs which colonizes mud flats of lower salinity. A salinity barrier of about 70 $^{o}/oo$ separates the gastropod habitats from the zones of growing mats. Under reduced salinity, the snails are able to destroy the microbial mats completely. (b) Decreasing regularity of flooding in the microbial mat environment of Mellum Island excludes intertidal deformative burrowers such as cockles and lugworms. However, locally the mats are pierced by numerous dwelling traces. These stem from small polychaetes and amphipod crustaceans which are able to spread over the intertidal-supratidal boundary and settle up to the MHWS-level.

Biogeochemical barriers are oxygen depletion within the sediments, high ammonia and sulfide contents, which generate through bacterial break-down of organic matter. Within the highly productive mats of *Microcoleus chthonoplastes* on Mellum Island, dwelling traces of marine polychaetes and amphipod crustaceans disappear due to these conditions. The name of the mat-forming species, *Microcoleus chthonoplastes*, indicates its capacity to form "soils" (Greek chthonos). While lithology is not altered, the presence of *Microcoleus* mats leads to a habitat change which excludes trace-making "arenophile" invertebrate species and favors "chthonophile" species which do not leave traces.

Stromatolitic microstructures studied in rock specimens were interpreted using modern analogs: Microcolumnar buildups in Precambrian stromatolites, ooids and oncoids were compared with those of modern microbial mats. The nodular to biolaminoid facies type found in the Gavish Sabkha was suggested to be an analog to the Plattendolomite facies of Permian Zechstein, North Poland. Studies of the Lower Jurassic ironstone of Lorraine clearly indicate that fungi have been involved in the formation of stromatolites, ooids and oncoids.

In conclusion, the comparative study of microstructures in microbial mats and stromatolites reveals a better understanding in both fields. In many cases, it was geology which first revealed the similarity of recent forms to those ancient ones and consequently encouraged research into them.

CONTENTS

PART I

LAYERED SEDIMENT ACCRETION BY MICROBES

- INTRODUCTORY REMARKS ON TERMS AND PROBLEMS -

"The name stromatolite relates to carbonate formations with a fine, more or less flat laminated structure, in contrast to the concentric formation of oolites ... In a certain sense there is also a transition between ooid and stromatoid. The transition from ooid, polyooid, ooid-bag to stromatoid denotes the increasing independance of the thin laminated structures from a center point."
(E. KALKOWSKY, 1908)

1. TERMS IN USE

Here we will comment on terms relating to sedimentary structures and products attributable to the activity of microorganisms (Table 1) with the aim of demonstrating that these terms cover one and the same genetic background. The general lack of definitions of terms in use led us also to these introductory remarks on the terminology of microbially generated sedimentary structures.

1. 1. Stromatolites and subsequent terms

KALKOWSKY (1908) used the term stromatolite to refer to layered patterns in rocks, produced by organisms which "have been so small that only the structure of their aggregates is preserved ..." (translated by KRUMBEIN, 1983). Subsequently, this visionary suggestion was more and more evidenced by refined methods of micropaleontology merging into paleomicrobiology (BARGHOORN & TYLER, 1965; KNOLL & AWRAMIK, 1983), namely that microorganisms were the main framework builders of stromatolites.

The term Spongiostromata was introduced by PIA (1927) to characterize fossil crustose growth structures. According to PIA, the Spongiostromata include stromatolites as well as oncolites, and his theory of origin was carbonate precipitation by crustose algae.

TABLE 1. Terms relating to microbially generated layered structures and particles

I.	LAYERED STRUCTURES FOSSIL AND RECENT: SYNONYMOUS TERMS	Stromatolites	(KALKOWSKY, 1908)
		Spongiostromata	(PIA, 1927)
		Algal sediments	(BLACK, 1933)
		Cryptalgal fabrics	(AITKEN, 1967)
		Algal mats	(GOLUBIC, 1976)
		Blue-green algal bioherms	(RICHTER et al., 1979)
		Microbial mats	(BROCK, 1976; KRUMBEIN, 1986)
		Growth bedding	(PETTIJOHN & POTTER, 1964)
II.	FABRICS WITHOUT DIRECT EVIDENCE OF MICROORGANISMS	Fenestral fabrics	(TEBBUTT et al. 1965)
		Thrombolitic fabrics	(AITKEN, 1967)
III.	PARTICLES	Oncoids	(HEIM, 1916)
		Ooids	(KALKOWSKY, 1908)

Subsequent studies of crustose algae in modern shallow subtidal and intertidal environments of the tropics and subtropics have supported PIA's idea that calcareous algae have built stromatolites. Accordingly, terms created to designate modern analogs of stromatolites were "algal sediments" or "algal mats". Further terms used are "cryptalgal fabrics" and blue-green algal bioherms since "blue-green algae" were observed to be most commonly involved in stromatolite formation.

The term "blue-green algae" is the traditional botanical assignment. However, taxonomically they are not algae but gram-negatively reacting, photosynthetic bacteria. Thus the taxonomically revised name of the group is now "cyanobacteria" (STANIER & COHEN-BAZIRE, 1977; KRUMBEIN, 1979c; RIPPKA et al., 1979).

However, we should avoid the term "cyanobacterial mats" to designate modern analogs of stromatolites for two reasons:

1. Although many stromatolites are in fact produced via photosynthetic activity of cyanobacteria, it seems important to stress that wavy laminated rock structures are not exclusively produced by "blue-greens". These structures can also originate from fungi and chemoorganotrophic bacteria (DAHANAYAKE & KRUMBEIN, 1985; DAHANAYAKE et al., 1985; DANIELLI & EDINTON, 1983; DEXTER-DYER et al., 1984; GYGI, 1981; KRETZSCHMAR, 1982; KRUMBEIN, 1983).

2. In the light of studies on modern laminated mats which display very complex biocoenotic systems including numerous groups of microbes and numerous metabolic pathways, it is assumed that stromatolites were produced by diverse microecosystems rather than by "monocultures". Cyanobacterial and fungal components are often well preserved in stromatolites due to their extracellular sheaths, envelopes and capsules, while other associated phototrophs and anaerobic heterotrophs are not in evidence (AWRAMIK et al., 1978; KNOLL & AWRAMIK, 1983). These, however, regulate the "physicochemistry" of a mat system and thus are fundamental. The biochemical activity of the associated bacteria is particularly important for the trapping and precipitation of minerals e. g. calcium carbonate, magnesium, copper, iron, manganese salts (KITANO et al., 1969; MITTERER, 1972; FRIEDMAN et al., 1973; KRUMBEIN, 1979a, b; WILSON et al., 1980; FERGUSON & BURNE, 1981; NOVITSKY, 1983; LUCAS & PREVOT, 1984; ECCLESTON et al., 1985; WESTBROEK et al., 1985), which often occur in association with stromatolites. Hence cyanobacteria and fungi are considered to be the main producers of organic substrate which support the succession and subsequent biochemical activity of other bacteria.

Accordingly the term "microbial mat" is preferentially used today to denote modern analogs of stromatolites (BROCK, 1976; KRUMBEIN et al., 1979; BAULD, 1984; COHEN et al., 1984). In their unconsolidated state, microbial mats of varying composition are also termed "potential stromatolites" (KRUMBEIN, 1983). A satisfactory definition of microbial mats has been given recently by KRUMBEIN (1986a).

To finish the list of terms associated with stromatolites and their modern analogs we refer to the atlas of primary sedimentary structures of PETTIJOHN & POTTER (1964), who included stromatolites inasmuch as they are "a type of growth bedding".

1. 2. Specific fabrics without direct evidence of microorganisms

Upon decay, sediments can be devoid of microbial cell remains but specific patterns such as fenestral and thrombolitic fabrics can indicate sediment accretion by microbes.

Fenestral fabrics in laminated microbial mats commonly generate from gas bubble formation, shrinkage and dessication (MONTY, 1976). The term "fenestra" was suggested by TEBBUTT et al. (1965) for a "primary or penecontemporaneous gap in rock frame work, larger than grain-supported interstices". Fenestrae were found within laminae of unicellular cyanobacteria which possess usually a great plasticity due to large quantities of gel around cell colonies. If in stratified mat systems, the gel-supported laminae are sandwiched between laminae built of filamentous microorganisms, the voids are elongated, follow the general bedding plane and describe a laminoid pattern (LF-A-type; MÜLLER JUNGBLUTH & TOSCHEK, 1969). On the other hand, more extensive layers dominated by unicellular organisms and their extracellular slimes can also show an irregular arrangement of fenestrae (LF-B-type). Laminated patterns, sedimentary augen structures and lensoids as well as the formation of oncoids and ooids in situ can be derived physically according to the law of pattern formation in laminae of different viscosities (D'ARCY THOMPSON, 1984).

Thrombolitic fabrics (AITKEN, 1967) in microbially produced sediments are due to irregular distribution of decaying dead colonies, intergrowing colonies or internal dissolution of mineral precipitates around colonies of microorganisms (MONTY, 1976).

1. 3. Biolaminated particles

The name oncoid was suggested by HEIM (1916) for spheroidal particles with non-concentric succession of more or less concentric laminae (FLÜGEL, 1982). PIA (1927) regarded them as a subgroup of the Spongiostromata, and his theory of carbonate precipitation by algae (see above) is still in use today, while HEIM's suggestion was that the formation of oncoids would be due to the "aggressive activity of bacterial colonies". Oncoids appear in both the fossil record and modern environments together with microbial mats. Cyanobacteria are commonly involved; we should, however, consider whether mineral precipitates around nuclei (bioclasts, microbial clots and lumps, lithoclasts) or empty spaces would be a better indication of "aggressive bacterial activity" (i. e. bacterial decay of the organic substrate).

Whether or not ooids are of biogenic origin is still a matter for controversy. The name was suggested by KALKOWSKY for more or less spherical or ellipsoidal grains with uniform, concentric laminae coating a nucleus. The use of the term ooid is rather complicated since it is understood to include at least two different kinds of origin (FLÜGEL, 1982). Generally, ooids and oolites (rocks consisting of ooids; TEICHERT, 1970) are studied with the consensus that they originate in high-energy environments. However, several modern mat environments show how it becomes possible to obtain laminated particles by the interaction of microbial communities with a physical and chemical environment.

LUDWIG & THEOBALD (1852) observed the formation of concentrically laminated coated grains in the thermal waters of Bad Nauheim which were called "Erbsensteine" i. e. pisoids, the terms ooid (KALKOWSKY, 1908) and oncoid (HEIM, 1916) being unknown at that time. The authors noted cyanobacteria- and diatom-dominated microbial mats in an open-air thermal water course and recognized the formation of coated grains around gas bubbles, metabolically derived from the mats. Furthermore, the authors noted that in fall the mat was degrading, which resulted in the release of the "Erbensteine", and their deposition downstream in sandy depressions as pisolites.

These, as well as various other studies, imply the existence of intimate genetic relationships between low energy environments and the formation of ooids during partition of microbial communities (WALTHER, 1885, ROTHPLETZ, 1892, GIESENHAGEN, 1922; SIMONE, 1981; FLÜGEL, 1982; MITTERER, 1971). FABRICIUS (1977), when studying the ultrastructure of Bahama ooids, noticed epilithic coatings of various particles with slimes of microorganisms, immediate precipitation of aragonite within the coatings and finally the genesis of concentrically laminated grains. He concluded that microbial partition in the genesis of coated grains is of primary importance, while oversaturation of the water with calcite, agitation and even nuclei supply are of secondary importance.

The recent finding of microbial mats building real domal stromatolites within the Bahama Bank environment strengthens the argument of the initiation of ooid formation within microbial mats also for the Bahama Bank ooid shoals.

In summary, we propose that the genetic definition of the term ooid should imply biogenicity rather than abiogenicity. We propose further that the term ooid should be placed into the genetically linked sequence of laminated sedimentary bodies and strata which form under the participation of microorganisms (see KALKOWSKY, 1908). The term ooid

TABLE 2. Classification of biolaminated deposits and particles (modified after DAHANAYAKE & KRUMBEIN, 1986)

	Laminated structures		Laminated particles			
Criteria	Single structure	Assemblage	Single particle	Assemblage	Single particle	Assemblage
GENESIS						
Biogenic	Stromatoid	Stromatolite	Ooid	Oolite	Oncoid	Oncolite
Abiogenic	Stromatoloid	Stromatoloid rock	Ooloid (*)	Ooloid rock	Oncoloid (**)	Oncoloid rock
MORPHOLOGY	Tabular, domed or columnar		Regular rounded		Irregular rounded	
LAMINATION	Planar to conical		Concentric continuous		Concentric discontinous	

(*) For ooids/ooloids larger than 2 mm in diameter the terms pisoid/pisoloid (pisolite/pisoloid rock) may be used

(**) For oncoids/oncoloids less than 2 mm in diameter the terms microoncoid/microoncoloid (microoncolite/microoncoloid rock) may be used

would then define a regularly concentric, the term oncoid an irregularly concentric coated grain. Finally the term stromatoid would include more or less stratiform lamina types (L_h- and L_v-laminae as described in this volume), hemispheroid structures (like the domal LLH- and separate vertically stacked SH-types classified by LOGAN et al., 1964; see also BATHURST, 1971). Logically, rocks composed of ooids (either authochthonous or allochthonous) are then oolites, of oncoids are oncolites and of stromatoids are stromatolites (Table 2).

Following the suggestions of BUICK et al. (1981) and BUICK (1984) we propose to avoid the above-mentioned terms if laminated deposits or particles are clearly abiogenic (e. g. geyserites) or where the biological origin is not unequivocal. In this case the said authors suggested the suffix "-oloid" (OEHLER, 1972) which means a "stromatolite-like" (ooid-like, oncoid-like) appearance. We propose to follow this advice with respect to the whole sequence (Table 2).

2. THE PROBLEM OF VERSATILITY

Fig. 1 has been constructed to illustrate the problem of versatility in microbial communities which may leave a lasting record in ancient sediments. Laminated or laminoid structures can derive from community types dominated by different major taxa (cyanobacteria, fungi or chemoorganotrophic bacteria; Fig. 1A). The wealth of different taxa solely in the group of cyanobacteria has to be considered. There are filamentous cyanobacteria, for example, which arrange themselves either horizontally, planarly or radially and leave behind lamina oriented either stratiformally (concordant to bedding planes) or upwardly convex.

Furthermore, experience of present-day microbial mat environments allows us to state that life strategies of the microbes concerned are so varied that they are able to develop over substrate of varying consistency. Presumably water is available permanently or periodically. It is therefore inadvisable to restrict the formation of stromatolites to any one particular environment which would imply in general their facies-indicative role (Fig. 1B). In the light of the great variability of lamina-forming microbes, it is essential to look carefully at other available sedimentological and paleontological information as far as facies are concerned. Stromatolites are neither necessarily intertidal nor always depth-dependent (JANNASCH & WIRSEN, 1981; MONTY, 1977; MASSARI, 1980). Dominance structures in microbial mats may be mainly environmentally induced. The biological growth habit of each single taxa in turn influences the morphogenesis. The process of burial and re-establishment is also important. Sedimentation is often involved, and microbial mats act as sticky fly papers (SHINN, 1983) which capture and bind allochthonous sediment particles. Growth of the microbiogenic sequence is also made possible without sedimentation as migrating organisms override others in order to find the most favorable environment (e. g. by phototaxis). This sort of self-burial is typical and can be seen in both lacustrine and marine low-energy environments.

Burial gives rise to various kinds of penecontemporaneous processes within the organic substrate (Fig. 1C): Micromilieus with varying degrees of bacterial decay occur; some polysaccharide chains and complexes are more recalcitrant than others; degassing, gas bubble formation and diffusion, shrinkage, dissolution and precipitation take place. The different plasticity of substrates and their behaviour due to compaction has to be considered. Faunal influence complicates the overall situation in so far as feeding and excretion lead to disintegration, and the spreading of microbial colonies may support the formation of thrombolitic fabrics and intraclasts.

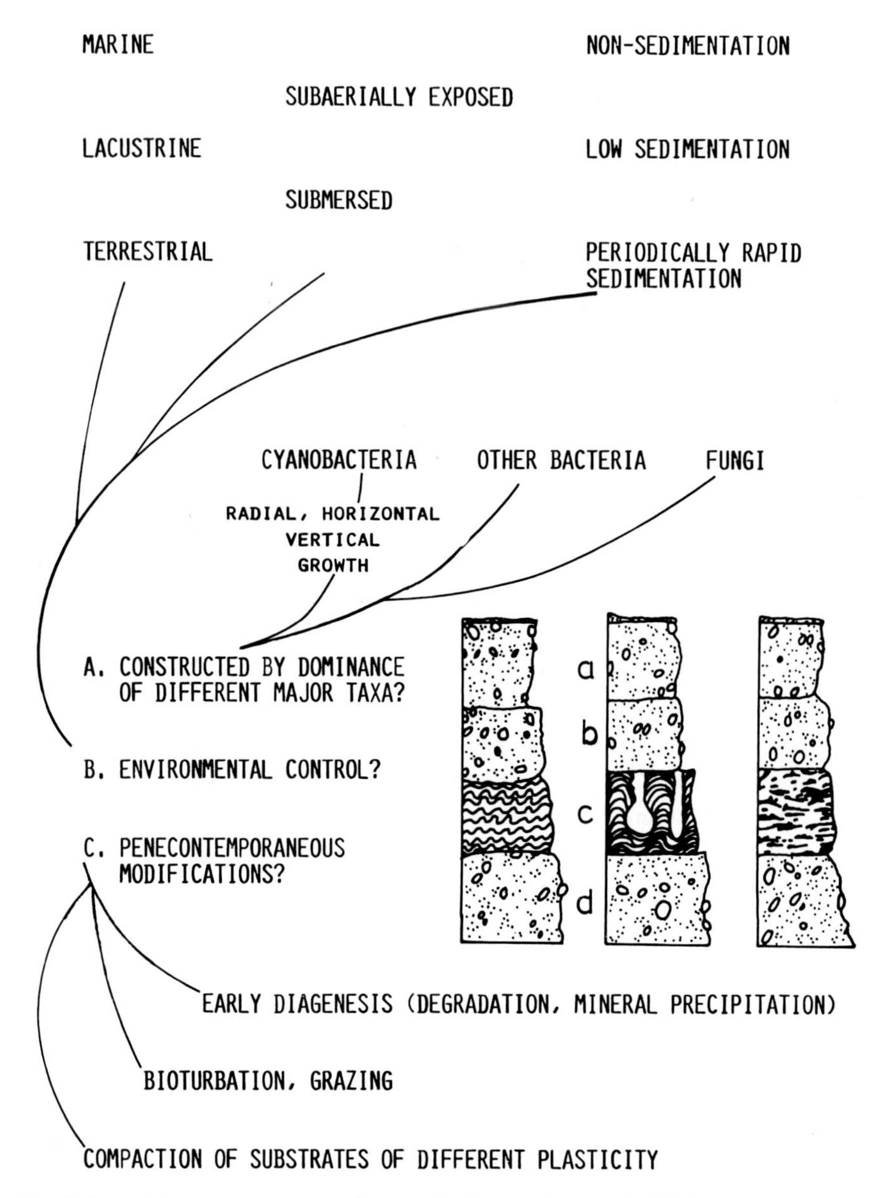

Fig. 1. Schematic representation relating stromatolitic structures and parameters possibly involved in their formation and early diagenesis. The concern of this scheme is to demonstrate that the appearance of a stromatolitic structure is indicative neither of one single phylum nor of one single environment.

The unifying principle within the complexity of stromatolitic fabrics is that they are products of microbes which by their morphology, physiology and arrangement in time and space interact with a physical and chemical environment to produce a laminated pattern (KRUMBEIN, 1983). This basic definition is irrespective of the existence of specific growth patterns (biostromate or biohermal buildups, laminated particles, and laminoid fenestral fabrics) which may be explained by biotopic and microbiocoenotic as well as by physical modifications.

A key to the recognition of biotopic and biocoenotic characteristics encoded within sedimentary structures is the study of microbial mats in modern environments. A limitation of the actualistic approach to stromatolites may be that many present-day depositional and ecological conditions do not function at the same rate and level of efficiency as in the past (REINECK & SINGH, 1980; KNOLL, 1985a). This consequently may explain why well developed thick stromatolitic sequences are less developed in the present than in the past. However, though restricted to the present-day extension of shelf flats and some special terrestrial, lacustrine and deep sea environments, present-day microbial mats witness clearly the constancy of the biolaminite tradition. The following chapters deal with stromatolite environments in the peritidal zone which apparently include types of potential stromatolites that always had world-wide distribution. Two of our main study areas are in the semi-arid tropics and are part of the desert coast adjacent to the Gulf of Aqaba graben system (Sinai Peninsula) while the others are supratidal flats of offshore embankments in the southern North Sea coastal region which is located in the temperate-humid zone. The studies describe microfacies types, bio- and ichnofabrics and modes of stratification which display depositional dynamics that interfere with microbial activity. Comparing the formative environments of these structures the important role played by climate and geomorphic relief becomes evident.

PART II

STROMATOLITE ENVIRONMENTS IN THE PERITIDAL ZONE

- MODERN EXAMPLES -

"Alle jene Gebiete, welche ich auf der geologischen Karte als 'Salzthon' ausgeschieden habe, sind nichts weiter als eingedampfte Lagunen und meerentblösster Strand." (JOHANNES WALTHER, 1888)

1. THE GAVISH SABKHA - A HYPERSALINE BACK-BARRIER SYSTEM (GULF OF AQABA, SINAI PENINSULA)

1. 1. Introduction

Facies is the product of specific depositional and biotopic conditions acting within a certain environment (TEICHERT, 1958). A specific biotopic condition in the Gavish Sabkha, a coastal environment in the arid tropics, is that the horizontal gradient of surface moisture is remarkably stable in the annual cycle. The availability of water is even more critical than the hypersaline environment if microbial mats are to flourish in the Gavish Sabkha (sabkha is a transliteration of the arabic term sabkhat or sebkat meaning salt swamp).

The Gavish Sabkha is a topographic low separated from the sea by bar-closing but in subsurface contact with the sea. Water loss by strong evaporation is constantly recharged by seepage seawater. Thus the system is provided with permanent shallow-water environments and moistened mud flats. These prevailing conditions can be interrupted although not always permanently changed by winter flashfloods.

The purpose of this chapter is threefold: (1) to document the effect of topographic moisture on the development of microbially produced sedimentary structures which represent analogs of conspicuous stromatolitic structures in the geological record (see for example KNOLL, 1985a), (2) to document the lithological and faunal framework which provide further information about the environment of deposition, (3) to interpret the mode of stratification of the sabkha deposits as the result of changes between long-lasting fair-weather conditions and short but catastrophic sheetflooding.

1. 2. Methods

Field work was carried out from July to October 1981 and February to March 1982. It focussed on the coring and documentation of undisturbed sediments, on the sampling of microbial mat material and benthic fauna, on measurements of physicochemical parameters and on the documentation of surface structures.

The Gavish Sabkha microbial mats were first studied by us in the early summer of 1978. At that time the permanently water-covered parts of the sabkha were floored with extraordinarily multilaminated communities (KRUMBEIN et al., 1979). Then two strong sheetfloods occurred, one at the end of 1979 and the other at the beginning of 1980 and the multilaminated microbial mats of the lagoonary basin were buried with loads of terrigenous sediments and died off. Our next studies were conducted in 1981, about one year after the floods. At that time not all of the benthic systems of the preflood period were reestablished although new initial stages had already developed. Thus, to interpret the fully developed stage of the multilaminated mat type which is repeatedly recorded in core segments, we adopt data from preflood studies (KRUMBEIN et al., 1979).

Data from seawater analyses and mineralogy of sabkha sediments presented here were also obtained during the preflood situation (GAVISH et al., 1985). In summary we present (1) data from preflood studies: ion concentrations in relation to the salinity of surface waters, mineralogy of surface sediments, permanently water-covered microbial mats (facies type 3), (2) data from postflood studies: topographic moisture and salinity gradients, sediment distribution, composition and standing crops of microbial communities which were already reestablished (facies type 1, 2 and 4), vertical profiles of redox potentials and pH in sediments, faunal distribution.

Undisturbed sediments were taken with plastic tubes (200/400 mm long, 50/70 mm diameter). Vertical sections were made with an electric knife, slices (45 x 45 mm) were separated and frozen under shock in liquid oxygen. They were subsequently dried in a dry-freezing apparatus and later hardened in an epoxy resin (Araldit F with hardener HY 905, REINECK, 1970). From the hardened specimens, thin sections were prepared and examined under a dissecting microscope. X-ray photographs were taken from the sliced core material as described by HAMBLIN (1962). Chips of the sliced core material were selected for scanning electron microscopy (SEM-Type Stereoscan 180, Cambridge Instruments). The material was fixed in glutaraldehyde (2 - 6 %) at appropriate osmolarities, then dehydrated in ethanol/H_2O dilution series and finally critical point dried. Subsequently the samples were sputtered with gold for SEM studies.

Sections of the sliced cores were selected for the study of the grain-size distribution. The sediments were wet-sieved into the fractions > 0.63 mm, 0.63 - 0.2 mm, 0.2 - 0.063 mm and < 0.063 mm.

Pore water was recovered for water analysis and sediment samples were carefully taken and subdivided for mineralogical, isotope and organic geochemistry work (FRIEDMAN & KRUMBEIN, 1985).

Physicochemical measurements (temperature, pH and redox potential) were carried out in freshly cored sediments. Temperature (air, water, sediments) was measured with a chrome/nickel thermoelement (Tastotherm, Gulton), pH and redox potential with electrodes (Ingold) on a millivoltmeter (Knick Portamess). Salinity was refractometrically measured (American Optical Instruments).

Microbial mat samples were examined under light and SEM microscopes. This work, as well as the physicochemical measurements, were carried out in collaboration with E. HOLTKAMP. Her data on community structures and pigment concentrations are included (HOLTKAMP, 1985).

The benthic fauna was collected qualitatively and semi-quantitatively with corers (190 mm diameter, 100 mm high). Samples were passed through a 0.5 mm sieve and fixed in 4 % formaldehyde. Specimens were sorted, identified with the help of specialists and counted. To study the burrowing behavior of salt beetles (*Bledius capra*) lab-cultured specimens were allowed to burrow in jelly seawater agar which was filled into glass tubes (150 mm high, 25 mm in diameter). To study the effect of grazing on microbial mats, gastropods (*Pirenella conica*) were allowed to graze on mat sections which were brought to the lab and treated with seawater at 50 o/oo salinity.

1. 3. Locality and previous work

The Gavish Sabkha is located in the southern coastal area of the Sinai Peninsula, opposite the Strait of Tiran, at 28° north latitude, 34°20' east longitude, and is approximately in 400 m distance from the Gulf of Aqaba (Figs. 2A and 2B). Shallow hypersaline surface water is widely surrounded by air-exposed saline flats (Fig. 2C). The general setting describes the Gavish Sabkha as a depression within an alluvial fan which spreads over the coastal plain between the Sinai Massive and the shoreline of the Gulf (Fig. 2B). Further north and south of the sabkha, the fan is crossed by major wadi conducts which on the occasion of flash floods transport terrestrial material into the coastal area, the Gavish Sabkha and the Gulf.

Sea-marginal sabkhas of the Sinai Peninsula were already mentioned by C. G. EHRENBERG (HEMPRICH & EHRENBERG, 1828) and J. WALTHER (1888). The Gavish Sabkha was first recognized by E. GAVISH, an Israelian geologist and geochemist. After the war in 1967, he began sedimentological, geochemical and hydrological investigations (GAVISH, 1974, 1980; FRIEDMAN & GAVISH, 1971; GAVISH et al., 1985). These pioneering studies were followed by studies of the microbial systems (KRUMBEIN et al., 1979; GERDES et al., 1985a; EHRLICH & DOR 1985). To honor the memory of E. GAVISH, who died in 1981, the comprehensive results of interdisciplinary research on sea-marginal sabkha environments using the example of the Gavish Sabkha were compiled (FRIEDMAN & KRUMBEIN, 1985).

1. 4. The physical environment

1. 4. 1. Geomorphic relief

GAVISH et al. (1985) proposed that the bedrock underlying the coastal bar consists of an uplifted reef complex and that the present softbottom sediments of the sabkha rest on an underlying backreef platform. Studies conducted by FRIEDMAN (1965, 1972) indicate that such uplift of reefs occurred about 2,000 - 4,000 years ago. Accordingly, GAVISH et al. (1985) suggest that the unique round shape of the sabkha and its location within the alluvial plain could be the result of a preexisting topographic low in the underlying reef platform which was subsequently uplifted.

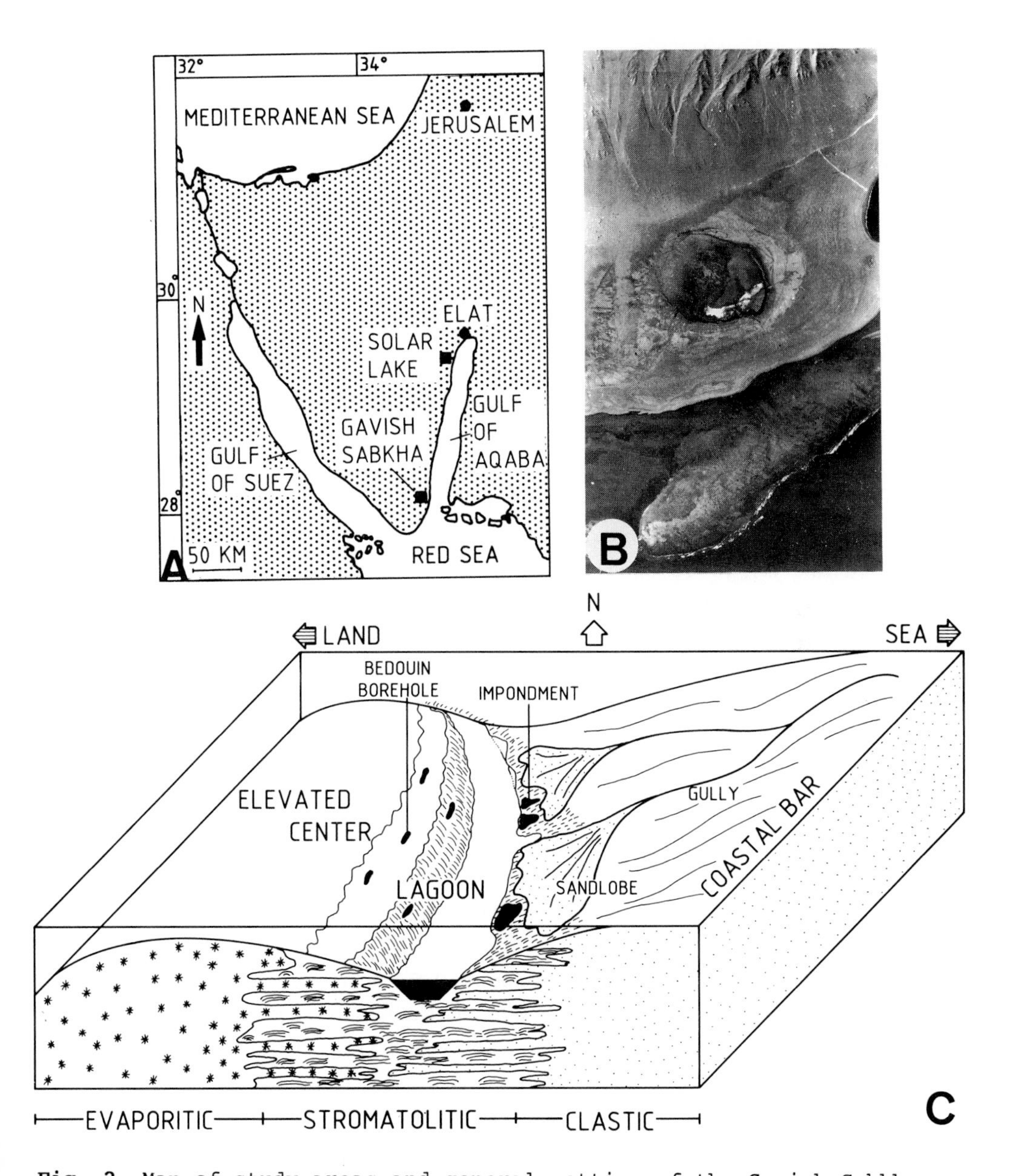

Fig. 2. Map of study areas and general setting of the Gavish Sabkha.

A) Sinai Peninsula with locations of the Gavish Sabkha and the Solar Lake along the Gulf of Aqaba. Modified after FRIEDMAN & KRUMBEIN, 1985.

B) A view toward W showing the round depression of the Gavish Sabkha at the shore of the Gulf of Aqaba. Fringing reefs are visible at the bottom, foot hills of the Sinai mountains at the top. After FRIEDMAN & KRUMBEIN, 1985.

C) Illustration of major geomorphic elements of the Gavish Sabkha: coastal bar slope, rims and basin of the lagoon, elevated center. Bar-directed siliciclastics and center-directed evaporites interfinger with microbial mats which form at the lower part.

The present-day environment forms a round depression about 500 m in diameter and is at its deepest part -1.80 m below m.s.l. (mean sea level). The central part of the depression is gently elevated and is surrounded by a concentric channel which is partially water-filled. The circular slopes rise gradually from the channel upwards to the alluvial plain and coastal bar facing.

Three major geomorphic elements of the Gavish Sabkha can be distinguished: The barrier slope, the lagoon and the center (Fig. 2C).

The barrier blocking the depression has been upheaved by coastline directed currents and waves. Seawater is replenished through subsurface conduits formed by carbonate cementation plates, fissures in the underlying bed rock and porous sediment infilling. The surface of the slope is covered by dry evaporite crusts. Several gullies cut through the slope face and merge at the lower end into sandlobes (Fig. 2C). Seawater springs rise at the sandlobe junctions. The gullies are generally 0.10 to 0.40 m deep and tend to meander. Gullies and sandlobes are the result of sheet floods (GAVISH, 1980). A wadi conduct runs at the coastal bar plateau.

The lagoon is part of the concentric channel (Fig. 2C). It forms a halfmoon-shaped basin with water depths of up to 0.60 m. The basin is surrounded by shallow and partially air-exposed flats where various impondments occur (about 50 mm deep and two to three meters in diameter) which are fed from seawater springs.

The elevation of the center is about 0.50 m above the water table of the lagoon. GAVISH proposed that gypsum accumulating below the surface caused the "swelling" of sediments and gradually upheaved the center. Aerial views show three sedimentary plains sloping gently to NNE (Fig. 2B): (1.) The topmost part, which is covered by dry evaporite crusts, (2.) the slope with its extremely gentle incline towards NNE and (3) the rim which is surrounded by the concentric channel. Surface moisture gradually increases toward the lagoon. Along the slope and the rim are several holes, excavated by fisher bedouins in order to collect brine and to harvest the precipitating potash and NaCl. During the winter months the sediments of the rim are submersed due to the slightly higher water table of the lagoon. In summer the water level retreats and leaves white gypsum crusts.

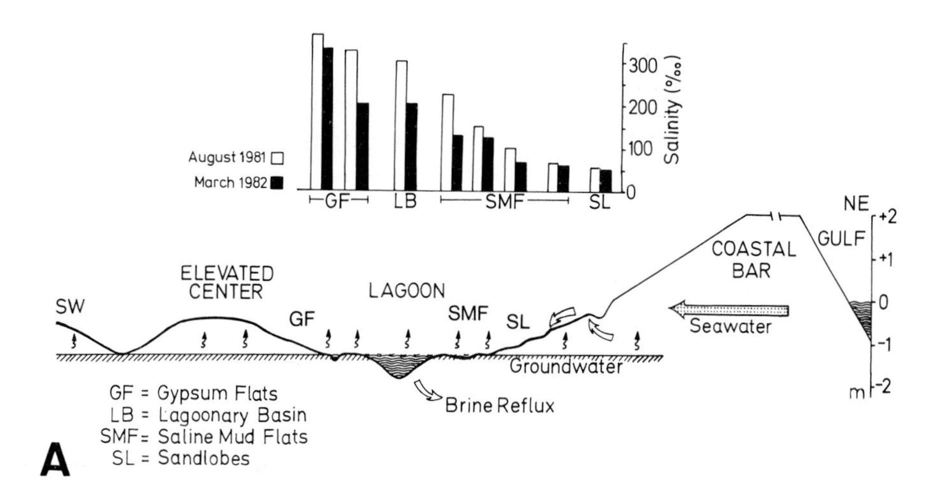

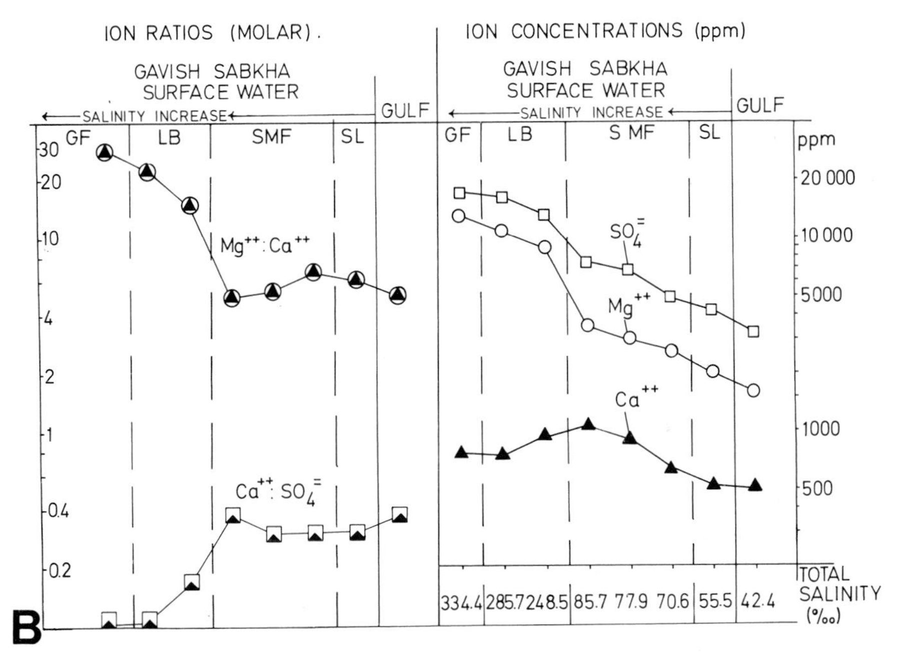

Fig. 3. Hydrology, salinity and seawater chemistry along a horizontal transect crossing several sub-environments of mat-formation: Sand lobes (SL), saline mud flats (SMF), lagoon (LB) and gypsum flats (GF).

A) Generalized horizontal transect showing the connection of the Gavish Sabkha with the Gulf. Hydrodynamic mechanisms are indicated by arrows: Seepage, evaporation and brine reflux. Salinity (upper diagram) increases towards the elevated center.

B) Ion concentrations (right) and ratios (left) in surface water including Gulf water (salinity values on abcissa right from GAVISH et. al, 1985; abbreviations refer to sub-environments listed in A). Decrease in Ca and change in ion ratios take place on saline mud flats bordering the lagoon where total salinity is about 85 ‰.

1. 4. 2. Hydrology

Two major effects on the hydrological system of the Sabkha have to be distinguished: (1) physical factors operating from the land and (2) physical factors operating from the sea. Those operating from the sea include seawater supply and water level changes within the sabkha due to tidal movements and weather effects (e. g. monsoon) outside. Onshore-directed wind drift, which occasionally causes strong wave attack against the coast, does not affect the bar-protected Gavish Sabkha. Thus the hydrological processes described below remain balanced even if high energy conditions occur in the adjacent Gulf.

Tidal influence. Gulf tides are semi-diurnal with a mean annual tidal range of 0.7 m. This tidal movement affects the water level of the Gavish Sabkha at a reduced rate with time delay. Diurnal movements of water levels of about 20 mm have been observed within the marginal pools and the basin of the lagoon. The operation of these tide-induced fluctuations is seen in the shifting of waterlines at the margins of the ponds and the basin. A sudden rise in the water table greater than normal (up to 10 cm) is common in winter. This phenomenon is not directly related to the tidal movement but to seasonal variations in the sea level of the Gulf as a response to climate conditions (monsoon; GAVISH et al. 1985). Both tidal- and climate-induced fluctuations of the water table do not lead to considerable disturbances of the microbial communities which colonize these margins. Some faunal elements, however, which are restricted to air-exposed wetland habitats, have to react by migrating according to the shifting water lines.

Seawater seepage. A constant supply of seawater is of greatest ecological significance. This is achieved by (i) the basin morphology of the Gavish Sabkha and (ii) the process of "evaporative pumping" (sensu HSÜ & SIEGENTHALER, 1969). The combination of seepage and evaporation is analogous with the boiling-pans formerly used by hill people to produce salt from seawater. They consisted of a pan with a high surface to volume ratio which was successively fed by low rates of seawater running through gently inclined pipes into the basin. The hydrological system of the Gavish Sabkha differs from this in two respects (i) the heat needed to evaporate the water generates from sun irradiation and (ii) the natural pan of the Gavish Sabkha consists widely of exposed sediments where the water table is below the surface. In these areas "evaporative pumping" operates (Fig. 3A).

The evaporation rate is about 4.6 m/year. Relative humidity averages 30 - 50 % with a mean annual air temperature of 26 ^{o}C and constantly high solar irradiation. The evaporation pumping mechanism operates both vertically and laterally. Percolating seepage seawater sinks into the sediments and elevates the water table, which is lowered subsequently by capillary evaporation. This stimulates upward movement of water within the phreatic zone by evaporative pumping and results in a constant supply of ions neccessary for mineral formation. The vertical capillary movement in turn leads to the lateral migration of fluids (PURSER, 1985).

The lateral movement induced by evaporative pumping may not operate where sabkha water tables rise landwards, as seen, for example, in the Abu Dhabi sabkha (PURSER, 1985). The downward inclination of the Gavish Sabkha, however, stimulates the lateral movement of interstitial water within the sedimentary system as well as the lateral supply of seepage seawater through permeable sediments of the marine bar. Carbonate plates embedded in the upheaved bar (Fig. 2C) probably serve as conduits for the seeping seawater. Their gentle inclination towards the sabkha can be seen at outcrops terrassing some of the gullies which cross the inward slope of the coastal bar.

Besides the precipitation of evaporite minerals at the sediment-air interfaces by evaporative pumping, reflux of brines to the sea is proposed to be an important mechanism of mineral accumulation (mainly gypsum) in deeper layers (GAVISH et al., 1985). Local dolomitization has also been suggested as an outcome of brine reflux, although several other models have been proposed.

Salinity regimes. A horizontal salinity gradient, ranging from 50 to 340 $^{o}/oo$, is established along a transect which runs from the lower slope face of the bar towards the center of the sabkha (Fig. 3A).

The data in Fig. 3A were obtained from measurements of interstitial water in the gully and sandlobe sediments and of standing surface waters. Although the data represent only two annual states (summer 1981 and spring 1982), they may nonetheless indicate that the establishment of salinity zones is more or less stable throughout the year. This assumption was reinforced by GAVISH on several earlier visits (GAVISH, 1975, 1980; GAVISH et al., 1985, see also Fig. 3B).

Short-term oscillations of salinity occur at the immediate rim of the lagoon where wind can flush-over concentrated brine. This deviation in the overall stability of the salinity regime is reflected by a very conspicuous change in microbially produced structures (see section 1.6.3).

Ion concentrations in surface waters. Fig. 3B shows a recalculation of GAVISH's data (GAVISH et al., 1985) concerning concentrations of dissolved calcium and sulfate in surface waters, Mg^{++} : Ca^{++} and Ca^{++} : $SO_4^{=}$ ratios. The data are correlated with salinity measurements obtained by GAVISH. These data are quite similar to our measurements taken four years later (Fig. 3A).

The increase of Ca^{++}-concentrations correlates to the salinity increase up to a solution barrier which lies at 80 to 90 $^{o}/oo$ within the saline mud flats (Fig. 3B). Further increase of salinity is accompanied by the decrease of Ca^{++}-concentrations in the surface water, accordingly the $Ca^{++}/SO_4^{=}$ ratio decreases, since sulfate takes over. The Mg^{++}/Ca^{++} ratio, on the other hand, shows a nearly linear increase. $CaSO_4$ is enriched in bulk sediments of the inner shoreline of the lagoon, while the sediments of the lagoonary basin where Ca^{++} is already depleted are enriched in high magnesium calcite (50 wt.%) and calcite (29 wt.%). As is shown in paragraph 1.6.4., these sediments are highly reduced so that bacterial sulfate reduction interferes with $CaSO_4$-precipitation.

Flashflood impacts. Annual rainfall in this area is 10 mm. Rainfall passing the system immediately is more or less negligible while run-off after rainfall in the adjacent mountains causes sheetfloods which transport sediment-laden freshwater. Such events have tremendous consequences for the hydrological and ecological conditions of the Gavish Sabkha. A strong sheetflood in October 1979 caused a rise in the water level of the Gavish Sabkha of about 60 cm. The steady state of the evaporation pumping system was completely disturbed. It took five months to regenerate (GAVISH et al., 1985).

1. 4. 3. Temperatures

A stable zonal temperature-gradient was not observed due to changing influences of wind, irradiation and evaporation operating in the daily cycle. In summer, air temperatures reach 45 °C and more and differences between day and night temperatures range between 20 and 30 °C. The nightly drop in air temperature is, however, less remarkable in shallow-water environments and below saltcrusts. Here, lower daily temperatures and a nightly drop ranging between 6 and 8 °C indicate the buffering capacity of shallow-water and evaporative crusts.

1. 5. Lithological framework

Data of GAVISH (GAVISH et al., 1985) were used to calculate the composition of surface sediments (0 to -5 cm) at different niveau levels. The results are presented in Fig. 4 in the course of a transect: The levels I to III indicate the three sedimentary plains of the center which gently dip to NNE (Fig. 4A). The level IV indicates the bottom of the lagoon which is the deepest part of the Gavish Sabkha. The levels V and VI describe different elevations of the coastal bar slope. The graded distribution of in-situ forming minerals described below demonstrates (1) the influence of geomorphology (i. e. the elevation of the relief above the groundwater table) (2) the effectiveness of evaporation pumping and seepage recharge of seawater prevailing in the annual cycle.

1. 5. 1. Evaporites

a) Halite (Fig. 4B). High amounts of halite (the relative frequency is over 80 %) accumulate at the elevated, permanently unflooded sediment surfaces (levels I and VI). This relative abundance of halite to other minerals indicates the total evaporation of upward moving marine water at the sediment-air interface. The thickness of the halite layer at the center (level I in Fig. 4A) averages 20 cm. It covers a gypsum layer with an average thickness of about 1 m (see paragraph c below). The relative frequency of halite decreases gradually with decreasing height, at level B the bulk volume is 25 - 50 % while at the lower levels III, IV and V it is negligible (Fig. 4B).

b) Anhydrite (Fig. 4C). Anhydrite occurs together with halite only in the upper dry crusts (levels I and VI), and there mainly in the lutitic fraction. The association of halite and anhydrite may indicate conditions of gypsum dehydration, while the frequency of anhydrite in

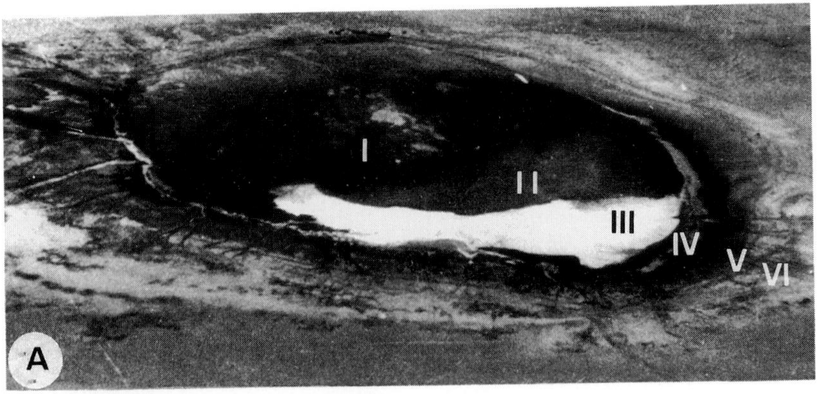

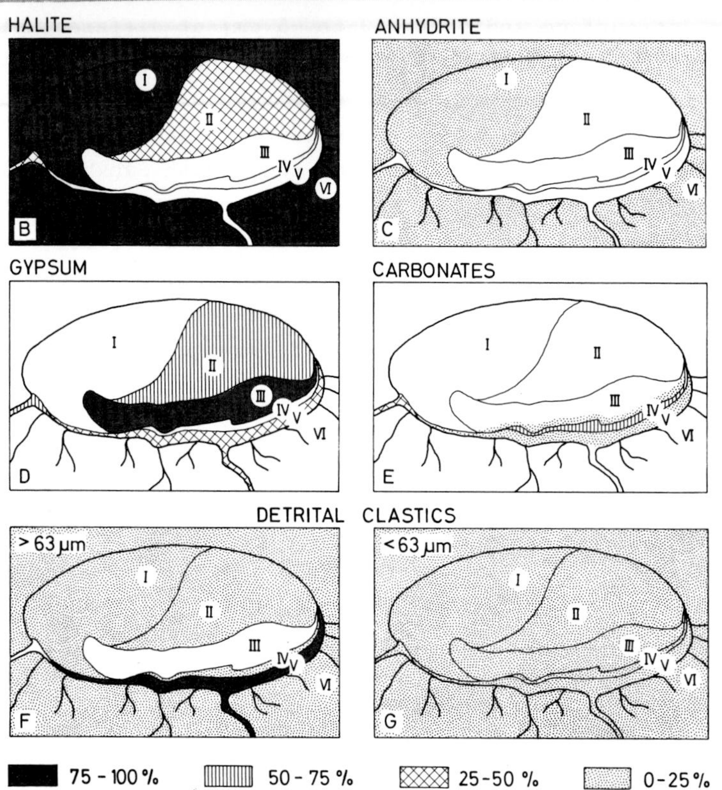

Fig. 4. Abundance of evaporites, carbonates and detrital clastics in surface sediments at different elevation levels of the Gavish Sabkha.

A) Aerial view from E towards W to show elevation levels: I - III: Upper, mediate and lower parts of the center (white area = gypsum), IV: Lagoon, V - VI: Lower and upper parts of the coastal bar slope.
B) Halite is in greatest abundance at the most elevated levels I and VI.
C) Anhydrite is generally associated with halite (levels I and VI).
D) Gypsum accumulates at lower levels (mainly level III, see also A).
E) Carbonates make up 50 to 70 wt.% of sediments of the lagoon (mg-calcite dominant, associated with calcite, aragonite, dolomite).
F) Sand-sized siliciclasts dominate at lower parts of the coastal bar.
G) Finer-grained detrital clastics generally occur at all levels.

the lutitic fraction indicates that it may be a primary precipitate. The hot and dry climate makes both processes possible.

c) Gypsum (Fig. 4D). The only zone where gypsum rarely contributes to the bulk volume of sediments is the lagoon (level IV in Fig. 4D). Here, biological sulfate reduction is most active. Within the sedimentary plain III in Fig. 4D, which is already part of the center but close to the lagoon, the bulk volume of gypsum averages nearly 100 %, thus indicating that the growth must be faster than the accumulation of clastic sediments deposited by wind. The relative frequency of gypsum decreases with increasing height while in turn halite becomes more and more dominant (compare the upward directed distribution from level III to level I in Fig. 4B). Below the halite crust of the elevated center, gypsum is the most authigenic mineral. With an average thickness of over 1 m, the layers reach well below the groundwater table surface (GAVISH et al., 1985).

1. 5. 2. Carbonates (Fig. 4E)

Carbonates are most frequent in the surface sediments of the lagoon and its margins. The major components are calcite and Mg-calcite (70 %), while dolomite averages about 5 %. Aragonite, which is an abundant carbonate component of the clastic sediments outside, is mostly a minor component or completely absent within the sabkha.

We will return to the question of in-situ carbonate accretion in this highly hypersaline milieu when referring to the development of biolaminated sediments and microbial habitats.

1. 5. 3. Detrital clastics

Terrigenous clastic sediments in bulk volumes of 0 - 25 % are mixed with the in-situ forming minerals. Coarser grains are without contact and are supported by matrices of gypsum or carbonate mud (wackstone-type). They represent fragments transported by northerly and southerly winds which blow several times in the year. The progradational tendency of the center sloping gently to NNE (indicated in Fig. 4A) may be mostly due to successive sediment infilling by southerly winds (PURSER, 1985; GAVISH et al., 1985).

Compacted surface layers of terrigenous clastics occur at the gully bottoms and sandlobes (level V in Fig. 4F). The grains are of low sphericity, which suggests primary provenance from weathering of Precambrian granitic rocks of the adjacent Sinai Massif (Fig. 5A and 5B).

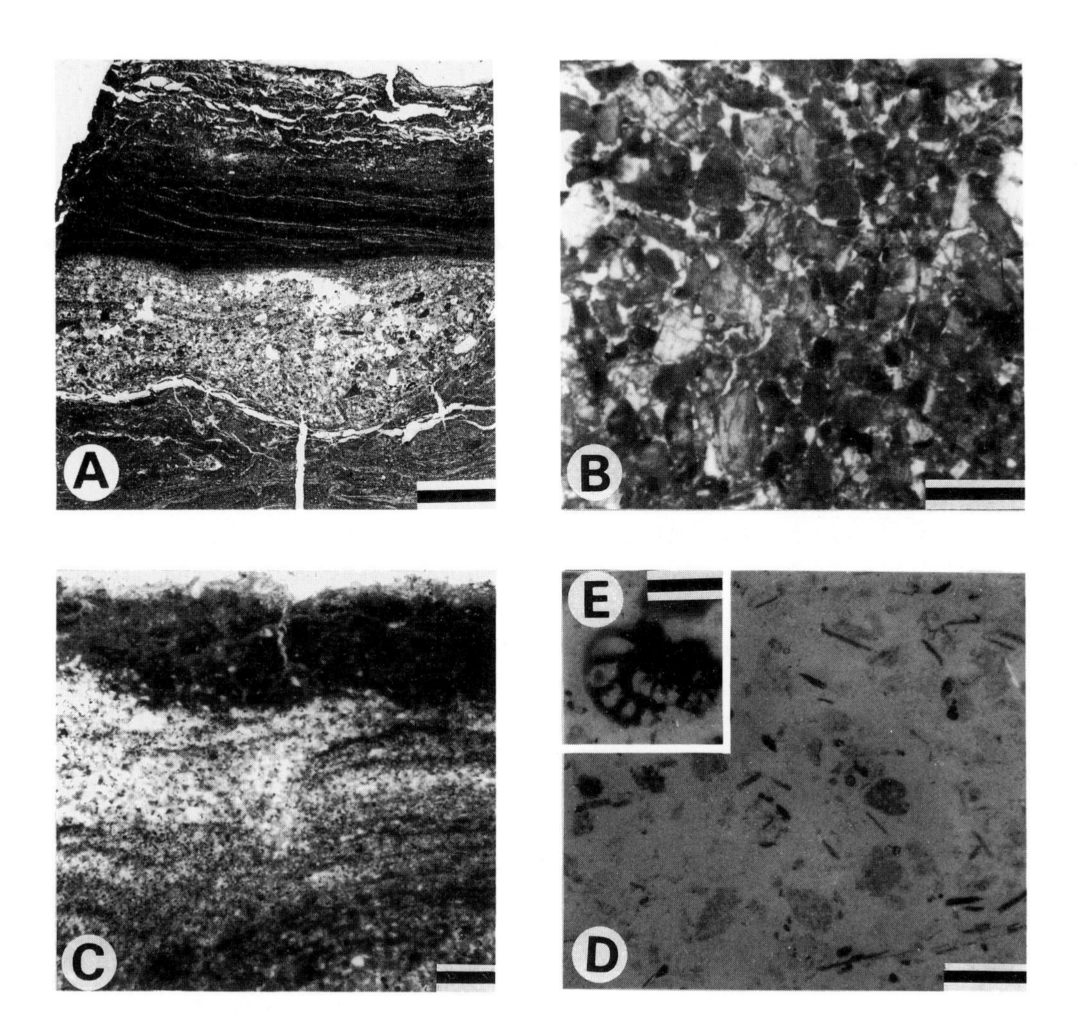

Fig. 5. Sheetflood deposits (thin sections from sediment cores).
A) Layer of grain-flow deposits between two microbial mat generations. Scale is 1 cm.
B) Texture of grain-flow deposits showing badly sorted and rounded particles. Scale is 500 µm.
C) Inverse grading of sheetflood sediments with water escape trace (or possibly escape trace of insect larva). Scale is 1 mm.
D) Sediments resulting from slow sinking deposition from suspension consist of silt, clay, plant debris and some coarser siliciclastic fragments. Scale is 250 µm.
E) Sheetfloods also transport seagrasses with epiphytic forams. Scale is 1 mm.

Since we carried out our field work approximately two years after the last strong sheetflood, we found all other sedimentary plains with the exception of the gully bottoms and sandlobes already re-covered by evaporites, carbonate mud or microbial mats. That strong sheetfloods cause debris to flow through a wadi system on top of the coastal bar and to spread over the whole of the lower-lying region is evidenced by several layers of coarse-grained material within the basin sediments of the lagoon.

Silt/clay fractions (Fig. 4G) derive from suspension clouds in freshwater which fill the depression during sheetfloods.

1. 5. 4. Internal fabrics of sheetflood deposits

a) Deposits resulting from debris flow: These are poorly sorted medium- to coarse-sized quartz sand concentrations which imply a gravity-induced lateral movement of debris loads in water (Figs. 5A and 5B). Inverse grading is recorded within several siliciclastic sequences (Fig. 5C). Around the circular slope of the sabkha depression the thickness of the debris flow deposits exceeds several dm. In the downslope direction, their thickness decreases. In sediment cores from the central basin, beds of debris flow range in size from a few mm to some cm.

b) Concentrations of silt, clay, plant debris, some foram skeletons and some coarser siliciclastic fragments (Fig. 5D): These concentrations result from slow sinking deposition from suspension. The interspersed coarser siliciclastic fragments without contact indicate transport in the clay-water fluid phase. The intermixed plant detritus of *Halophila* sp. stems from the mangrove swamps and lagoons to the north of the Gavish Sabkha in the order of tens of kilometers. Epiphytic forams (determined as *Sorites* sp. by L. HOTTINGER) are mechanically transported with seagrass leaves into the Gavish Sabkha (Fig. 5E).

Beds of silt and clay are rarely distributed around the circular slope but are characteristic of the sediments of the central basin where the clay-water fluid phase comes to rest. The deposits from suspension are a few mm to several cm thick.

1. 6. Stromatolitic facies types

1. 6. 1. The microbiota

Various species of procaryotic and eucaryotic microorganisms participate in forming biogenic structures in the Gavish Sabkha (Table 3). Since taxonomic determinations are still uncertain, we will refer only to the genera of the organisms found. Several of these genera are present with more than one species (e. g. *Oscillatoria*, *Spirulina*, *Synechococcus*, *Thiocapsa*, *Nitzschia*, *Navicula*).

TABLE 3: Microorganisms in the microbial mats of the Gavish Sabkha

I. Procaryotes

A. Unicellular cyanobacteria
Gloeothece, *Synechococcus*, *Johannesbaptistia*, *Gloeocapsa*, *Synechocystis*, *Myxosarcina*, *Pleurocapsa*, *Chroococcodiopsis*

B. Filamentous Cyanobacteria
Spirulina, *Oscillatoria*, LPP-forms[1]: *Microcoleus*, *Hydrocoleum*, *Phormidium*, *Lyngbya*, *Plectonema*, *Schizothrix*

C. Anoxyphotobacteria
Chromatium, *Thiocapsa*, *Ectothiorhodospira*, *Chloroflexus*

D. Chemolithoautotrophic bacteria
Thiobacillus, *Beggiatoa*, *Desulfovibrio*)[2]

E. Chemoorganotrophic bacteria
Pseudomonas, *Spirillum*, *Spirochaeta*, *Proteus*, *Desulfovibrio*, S^{o}-reducing and other taxa

II. Photosynthetic eucaryotes
Diatoms: *Mastogloia*, *Navicula*, *Amphora*, *Nitzschia*[3]

[1] "LPP"-grouping refers to Rippka et al. (1979). LPP stands for the genera *Lyngbya*, *Phormidium* and *Plectonema* which are representativs for structural and physiological characteristics of the LPP-Group. Microcoleus should be incorporated in LPP

[2] Physiological attributes mentioned are only partially significant

[3] Only frequent forms; for more details see Ehrlich & Dor (1985).

1. 6. 2. Major mat-structuring organisms

Types of primary producers which give the Gavish Sabkha mats their structure are (1) heavily ensheathed filamentous cyanobacteria (2) capsulated unicellular cyanobacteria with multiple fission (3) slime-ensheathed cyanobacteria with binary fission.

(1) The main ensheathed filamentous cyanobacteria present is *Microcoleus chthonoplastes* (Fig. 6A). It is a cosmopolitan species found in

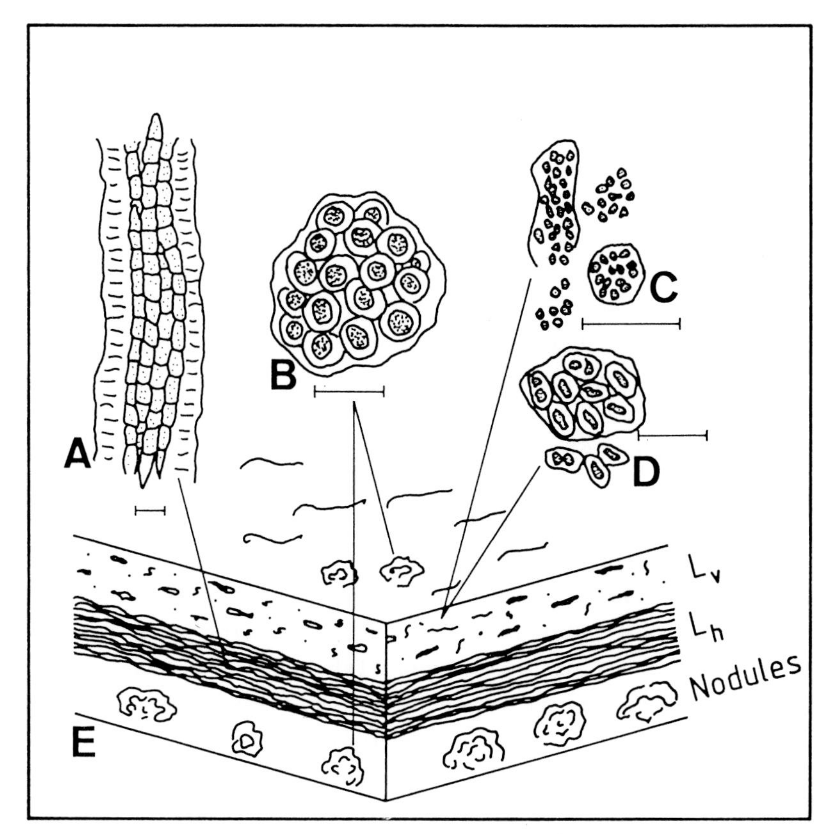

Fig. 6. Main mat-structuring microorganisms (modified after EHRLICH & DOR, 1985). Scale is 10 µm for all presentations.
A) Ensheathed filament bundles of *Microcoleus chthonoplastes*.
B) Pleurocapsalean cells encased in polysaccharid capsules.
C) *Synechocystis* sp.
D) *Gloeothece* sp. The latter both represent coccoid unicells in colloidal matrix.
E) Resulting depositional structures: Colonies of *M. chthonoplastes* produce horizontally oriented laminae (L_h), Pleurocapsalean colonies form cauliflower-shaped nodules, *Synechocystis* and *Gloeothece* contribute to porous, slime-enriched layers containing bubbles and some diagonally to vertically oriented filamentous organisms (L_v).

various environments e. g. Spencer Gulf, Australia (BAULD, 1984; SKYRING et al., 1983), Laguna Mormona, Mexico (STOLZ, 1983; MARGULIS et al., 1980), Solar Lake, Egypt (KRUMBEIN, 1978; COHEN, 1984), Farbstreifen-Sandwatt, southern North Sea coast (OERSTEDT, 1841; HOFFMANN, 1942; GERDES et al., 1985b, c; STAL, 1985; see also this part chapter 3). Multiple ensheathed filament bundles are typical of this species which are rarely if ever found in a diagonal or vertical arrangement. Microbial mats dominated by this species reveal a smooth and uniformily flat microtopography. Within vertical sections the mat appears as a horizontally layered, bedding plane concordant lamina which we call a L_h-lamina (Fig. 6E; see facies type 1 and 3 in paragraph 1.6.3).

(2) The most common capsulated unicellular cyanobacteria with multiple fission is *Pleurocapsa* sp. (formerly *Entophysalis* sp.). Species of Pleurocapsalean form cell colonies where each individual cell is encased by thick concentric lamellated sheaths (Fig. 6B). A crucial difference from *Microcoleus chthonoplastes* is that the coccoid colonies do not form flat and bedding-plane concordant mats but reveal discontinuous, more or less concentric structures. The microtopography of sediment surfaces which are colonized by Pleurocapsalean populations exhibits a pustular structure. Cauliflower-like nodules are also common at sediment surfaces (Fig. 6E; see also facies type 2 in paragraph 1.6.3).

(3) The most common slime-ensheathed unicellular cyanobacteria with binary fission are *Gloeothece* sp. and *Synechocystis* sp. (Figs. 6C and D). These organisms account for the vast production of polysaccharides. Sediments composed of or interwoven with these polysaccharides are sluggish and yoghurt-like if liquid (e. g. seawater) is maintained, due to the dispersal and partial dissolution of the mucilage. The cells of the organisms are irregularly arranged in the slime mass (MARTIN & WYATT, 1974). The species mentioned contribute predominantly to immense slime layers found in the Gavish Sabkha sediments. Since the slime-supported coccoid mats form vertically extended layers which differ from the flat and bedding-plane concordant L_h-laminae of the *Microcoleus* mats, we call them L_v-laminae (Fig. 6E; see also facies type 3 in paragraph 1.6.3).

These unicellular organisms increase their slime production in order to escape phototoxic conditions when they form the surface mats. Slime production is also stimulated by increase in salinity and temperature (CASTENHOLZ, 1984). All these conditions occur in the Gavish Sabkha. Photosynthetically active populations of other species (e. g. *Microcoleus chthonoplastes*) under the translucent mat benefit from the production of large quantities of gel since it is an ideal medium for the channelling of light.

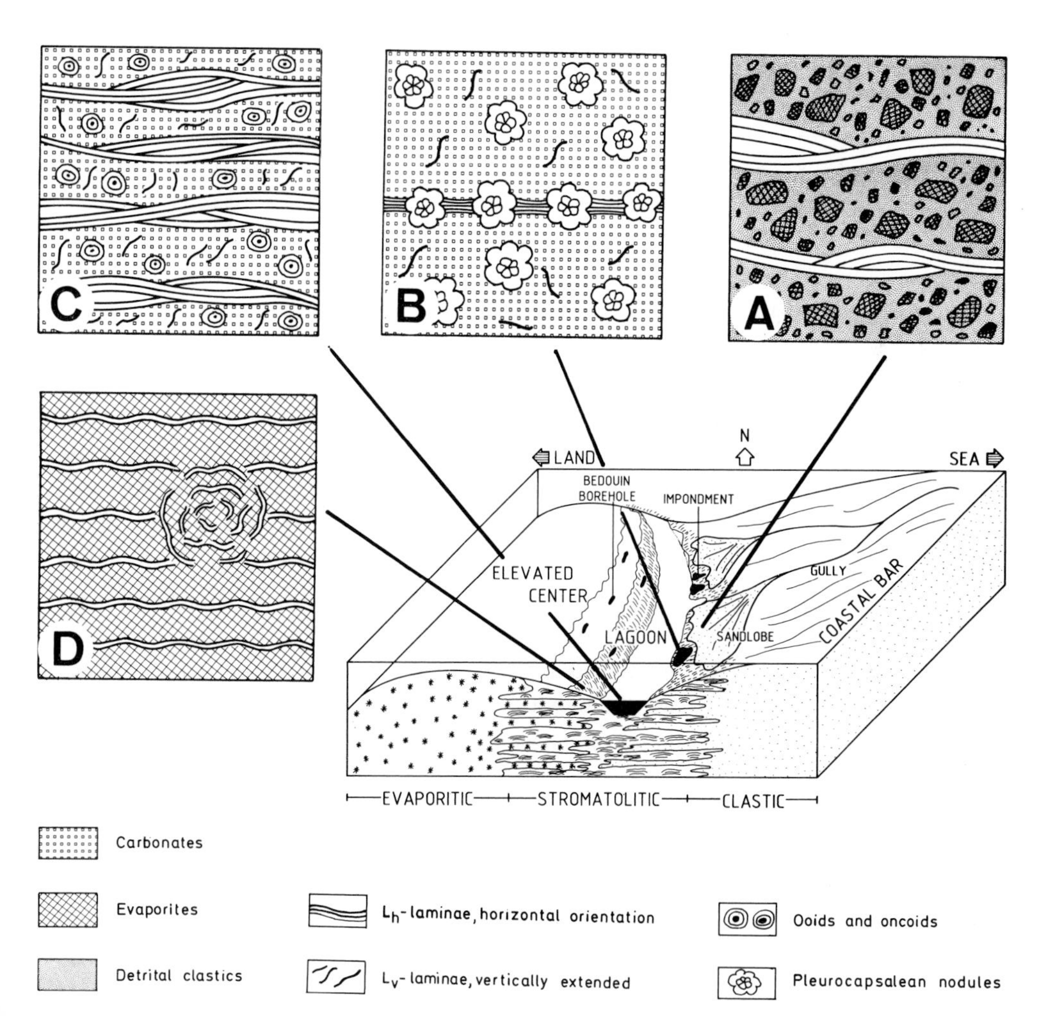

Fig. 7. Local distribution of stromatolitic facies types. The sequence A) to D) correlates to increasing salinity. The possibility of superficial water increases from A) to C) and decreases again in type D).
A) Siliciclastic biolaminites (gully bottoms and sandlobes).
B) Nodular to biolaminoid carbonates (saline mud flats).
C) Stromatolitic carbonates with ooids and oncoids (lagoon).
D) Biolaminated sulfate (gypsum flats).

1. 6. 3. Character and distribution of stromatolitic facies types

A supply of moisture to surface sediments where the initial growth of microbial mats occurs is essential for their development. In the Gavish Sabkha moisture at or close to the sedimentary surfaces is a function of topography. Since the topographic moisture gradient parallels other facies-relevant factors, e. g. salinity (Fig. 3) and in-situ mineral accumulation (Fig. 4), this gradient will be used as the variable to describe the lateral distribution of facies types. An informal overview of the lateral sequence is given in Fig. 7. Their individual patterns are described in the following sections in terms of community structures, environments, sedimentary textures and structures. Community structure is defined here as the visually observable product of species selection and dominance at a certain place.

1. Siliciclastic biolaminites (Fig. 8)

This facies is composed of quartz sand and interlayered biolaminites. The biolaminites characterize the L_h-type which is *Microcoleus*-dominated (Fig. 6A). Thickness of laminae differs from 50 to 500 µm. Where laminae are thin (about 50 µm), they form a monolayered mat. Thicker laminae comprise several generations of mat development, aided by sufficient surface moisture during periods of non-deposition (Fig. 8A).

Environments are the gully bottoms and sandlobes along the barrier slope (Fig. 7). Here the groundwater table runs -5 to -10 cm below surface. Vertical movements of the groundwater table between 2 - 5 cm depend on the external tidal movement in the Gulf. Sediment surfaces are permanently wetted by capillary movement of the groundwater (evaporative pumping). Thin evaporite crusts (mainly gypsum) form at sediment-air interfaces. During winter time, oscillations of the groundwater table lead occasionally to the inundation of the sedimentary surfaces. Salinity of interstitial water is 45 - 60 $^{o}/oo$. No extreme shifts occur during the annual cycle.

Sediments are mainly terrigenous and consist of badly sorted medium to gravel-sized siliciclastic sand. The grains are of low sphericity (Fig. 8B). These sediments originate from debris-flow.

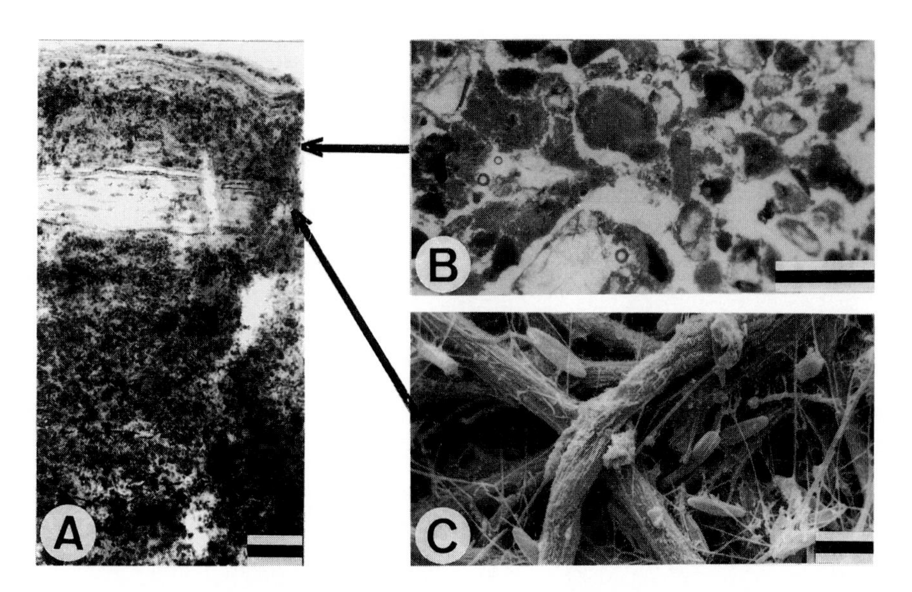

Fig. 8. Documentation of facies type 1: Siliciclastic biolaminites.
A) X-ray radiograph of a sediment core showing microbial mats in 3 cm depth and at the surface. Scale is 1 cm.
B) Close-up of siliciclastic sediments showing microbial coatings of sediment particles (center). Thin section. Scale is 1 mm.
C) Close-up of biolaminites showing members of the mat community: sheathed bundles of *Microcoleus chthonoplastes*, diatoms and filamentous sulfur bacteria. SEM-photography. Scale is 50 µm.

Community structure (Fig. 8C): The bulk of the laminae is made of ensheathed bundles of *Microcoleus chthonoplastes*. Few unicellular genera of cyanobacteria and diatoms are associated with the *Microcoleus* mat. The mat develops -2 to -5 mm below the water surface or underneath evaporite crusts. Both siliciclastic sediments and evaporite crusts support light-channelling for photosynthetic activity.

2. Nodular to biolaminoid carbonates (Fig. 9)

This facies type is characterized by granular, cauliflower-shaped microbial aggregates (nodules) embedded in wide-spaced biolaminoids (a term which defines a less significantly laminated build-up of biogenic sediments). The embedding material consists of intermixed calcite mud and mucilaginous polysaccharides. Various tubular relicts of empty sheaths of filamentous cyanobacteria are visible around the nodules (Fig. 9A). Transitional stages towards laminoid arrangements of smaller

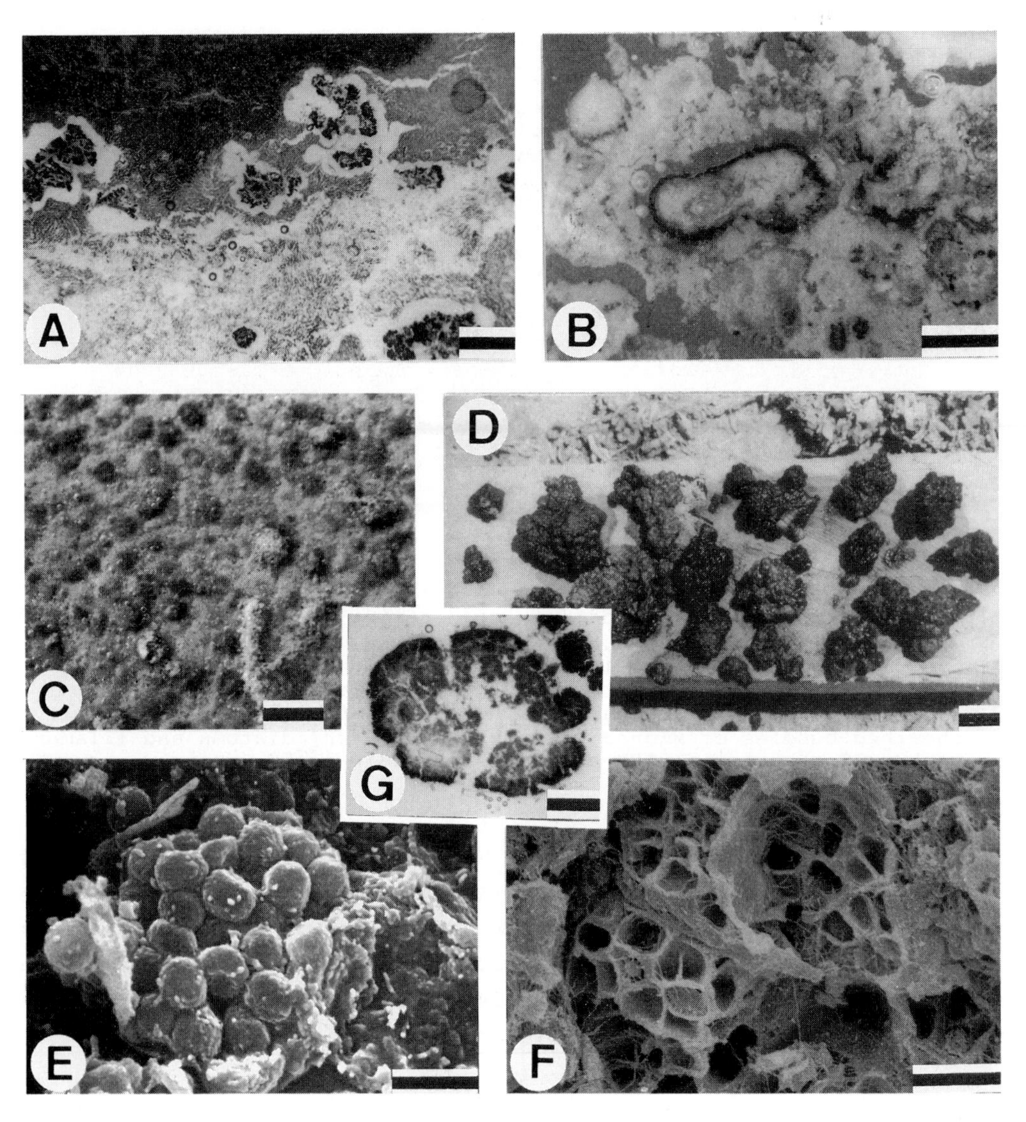

granulae and filaments can be seen (Fig. 9B). Surfaces of the mats exhibit a mammillate microtopography (Fig. 9C). The pattern of this facies type is stimulated by constant shifts of salinity and depth of water.

Environment. The nodule-biolaminoid facies type dominates saline mud flats just beyond and in between the sandlobes (Fig. 7). The saline mud flats border the lagoonary basin and are partially built over by the

Fig. 9. Documentation of facies type 2: Nodular to biolaminoid carbonates (saline mud flats at the lagoon's outer rim).
A) Intrasedimentary nodules embedded in carbonate mud. A filigrane meshwork of diagonally to vertically oriented filaments is visible. Dark sediments at top: reduced. Scale is 500 µm.
B) Intrasedimentary twin-nodule and biolaminoid structures (note wavy diagonal dark line in the right upper corner). Scale is 2 mm.
C) Mammillate surface microtopography, characteristic of the nodule-bearing zone. Scale is 1 cm.
D) Larger nodules sampled at the surface of small water-filled puddles. Scale is 2 cm.
E) SEM-photography of a colony of nodule-forming cyanobacteria (Pleurocapsalean). Scale is 3 µm.
F) SEM-photography of nodule compartments showing capsules of former cells. Note radial arrangement of compartments. Scale is 3 µm.
G) Dissected nodule showing radial arrangement of compartments, the empty center and iron-rich pigments around the cortex. Scale is 2 mm. Figs. A, B and G thin sections from sediment cores.

---

elevated sandlobe "bars" of coarse sediment which project from the gullies into the lagoon. Seepage seawater merges at the sandlobe junctions and feeds the gently downwards sloping mud flats with a trickling water film. Lower salinity at seawater springs between 50 - 70 $^{o}/oo$ is marked by a thick scum of benthic macroalgae (*Enteromorpha* sp.). Microbial mats do not develop here. The seepage water accumulates in shallow embayments and drains from these embayments through narrow passages into the central basin. Salinity increases with distance from the seawater springs from 70 to 180 $^{o}/oo$. Some very shallow embayments (maximum water cover 10 cm) are subject to short-term (commonly diurnal) changes of water cover and air exposure due to changing conditions of evaporation and wind velocities. Short-term salinity shifts were observed ranging from 70 to 150 $^{o}/oo$. HOLTKAMP (1985) recorded more extreme amplitudes ranging within a few minutes between 130 and 240 $^{o}/oo$. Other embayments with water levels between 20 and 40 cm remain water-filled during diurnal and annual cycles. The salinity ranges from 70 to 120 $^{o}/oo$. Flats around the embayments are usually air-exposed and covered by thin evaporation crusts. Salinities here can reach up to 180 $^{o}/oo$.

Sediments are mainly composed of fine-grained carbonates. Grain size analyses of surface sediments show 38 wt.% <2 µm, 27 wt.% 2 - 6.3 µm and 30 wt.% 6.3 - 20 µm. The sediment is mixed with mucilaginous polysaccharides and is of a sluggish, yoghurt-like appearance. At the water-air interface of the embayments, thin and fragile "calm skinned" evaporite crystals form and are immediately disintegrated by wind. The

Fig. 10. Documentation of facies type 3: Stromatolitic carbonates with ooids and oncoids (lagoonary basin).

A) Sequence of dark (L_h) and light laminae (L_v), growing independently from sediment transport, calcification being a penecontemporaneous process (note coatings of diagonally to vertically oriented filaments within the light layer). Scale is 1 mm.

B) Close-up of coated filaments within a L_v-lamina. SEM-photography. Scale is 5 µm.

C) Ooid embedded in calcified filamentous meshwork. Scale is 500 µm.

D) Cluster of coated grains within an extended light lamina. Scale is 500 µm.

E) String of coated grains captured by horizontally oriented filaments. Scale is 1 mm.

F) Eye-shaped lense, formed by a discontinuous L_h-lamina. Lenses are often the micro-environment of coated grain formation. Note rigid calcification below, within and above the dark lamina. Scale is 500 µm.

G) Ooid surrounded by various intraclasts (dark spots). Scale is 500 µm.

H) View of a biohermal build-up of a mat at the lagoon's margin. The partial destruction is due to slight wind surf. Scale is 5 cm.

Figs. A, C, D, E, F and G thin sections from sediment cores.

---

fragments sink down to the bottom and form a water-saturated soft gypsum mush. Some portions become diluted, others are embedded into the mucilaginous polysaccharides and get colonized by the microbial communities.

The community structure is dominated by unicellular cyanobacteria with multiple fission (Pleurocapsalean types; Figs. 9E, F). Other unicellular forms (predominantly the major slime producers *Gloeothece*, *Synechococcus* and *Synechocystis*) and filamentous cyanobacteria are also present. The Pleurocapsalean cyanobacteria form cauliflower-shaped colonies (Fig. 9D) which occur intrasedimentarily (Fig. 9A) and at the sediment surface (Fig. 9E). The compartments of these "nodules" when dissected show concentric growth around cavities (Fig. 9F). The yellow-brown iron-enriched pigment scytonemine colors the cortex of each colony compartment and especially the outer surface of the nodules (Fig. 9F). The granular surface microtopography of the nodules is maintained following division of the individual cells. With the lateral continuation of the saline mud flats into the lagoonary basin, the facies type changes to a more regularly spaced pattern (see next section).

3. Stromatolitic carbonates with ooids and oncoids (Fig. 10)

This facies type is characterized by regular interlayering of light and dark laminae (Fig. 10A). Both lamina types differ in the vertical

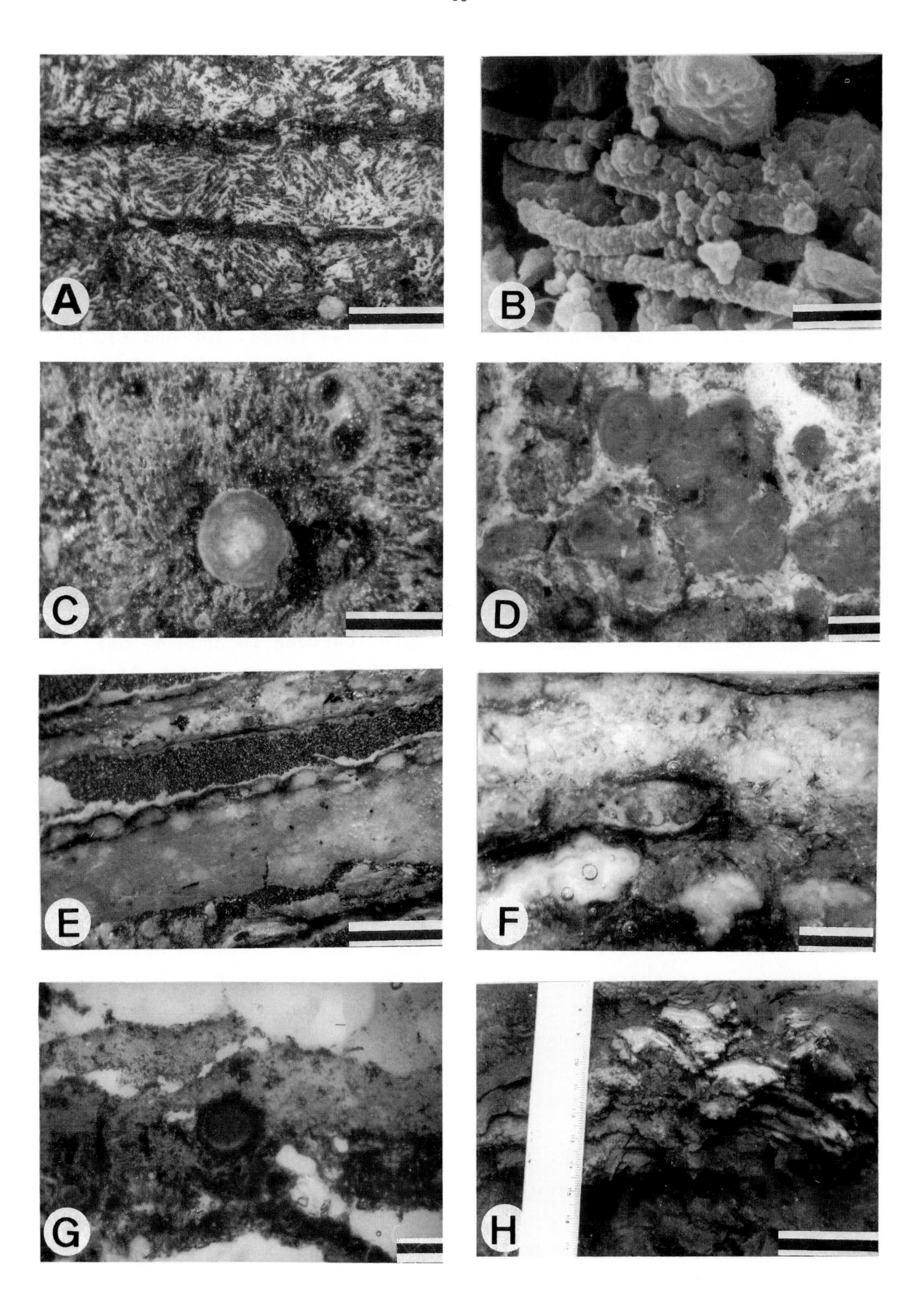
A
B
C
D
E
F
G
H

extension: The dark lamina is generally thin (100 to 200 µm), while the light lamina is up to ten times thicker. Various vertically to diagonally oriented filaments thread through the light layer, demonstrating active movements of unsheathed hormogonia and trichomes of filamentous cyanobacteria. These filaments frequently show calcite coatings (Fig. 10B). Associated with the light layers are also carbonate grains of varying diameters (Figs. 10C to 10F). Grains within thickening light layers show dispersed and sometimes clustered arrangements (Fig. 10D). Others appear like strings of pearls, captured by filaments of the dark laminae which form pockets or eye-shaped lenses (Figs. 10E and F). Some grains are regularly concentric (ooidal), others are irregularly shaped with transitions from oval to subangular forms (oncoidal). Vertical sections show that concentric lamination of these grains is common (Fig. 10G).

The <u>environment</u> is the lagoonary basin which is fed by seepage seawater and probably artesian discharge (wells) at the basin bottom. The 70-cm-deep center of the basin slopes shallowly upwards. The said facies type develops mainly along the shallow shelf area which is subject to seasonal changes of water depth (between 10 and 40 cm) and salinity ranging from 140 (winter) to 280 $^{o}/oo$ (summer; Fig. 3). These annual changes are performed in a long term, even pattern which is in contrast to the extreme short term irregularities of water supply and salinity in the shizohaline amphibic zone of the nodule-biolaminoid facies. The basin rim is subject to a slight surf caused by the wind. Under surf attack the microbial communities tend to form biohermal build-ups (Fig. 10H; see also Fig. 11A).

<u>Sediments</u>: Recalculation of GAVISH's data (GAVISH et al., 1985) shows bulk volumes of carbonate minerals, the most commom being high magnesium calcite averaging 50 wt.% and calcite (29 wt.%). Dolomite is found with 5 wt.%. The remaining portion consists of sand-sized (10 wt.%) and silt-sized quartz (6 wt.%).

<u>Communities</u>: The matrix of this biofacies type is primarily built of microbial biomass, mainly of cyanobacteria. Biomass accumulates under the evenness of shallow-water cover, giving rise to a multilaminated living unit which is clearly of stromatolitic appearance. The dark laminae are created by the predomination of heavily ensheathed filamentous cyanobacteria, mainly *Microcoleus chthonoplastes* (L_h-type). The light laminae (L_v-type) are gel-supported and produced by unicellular

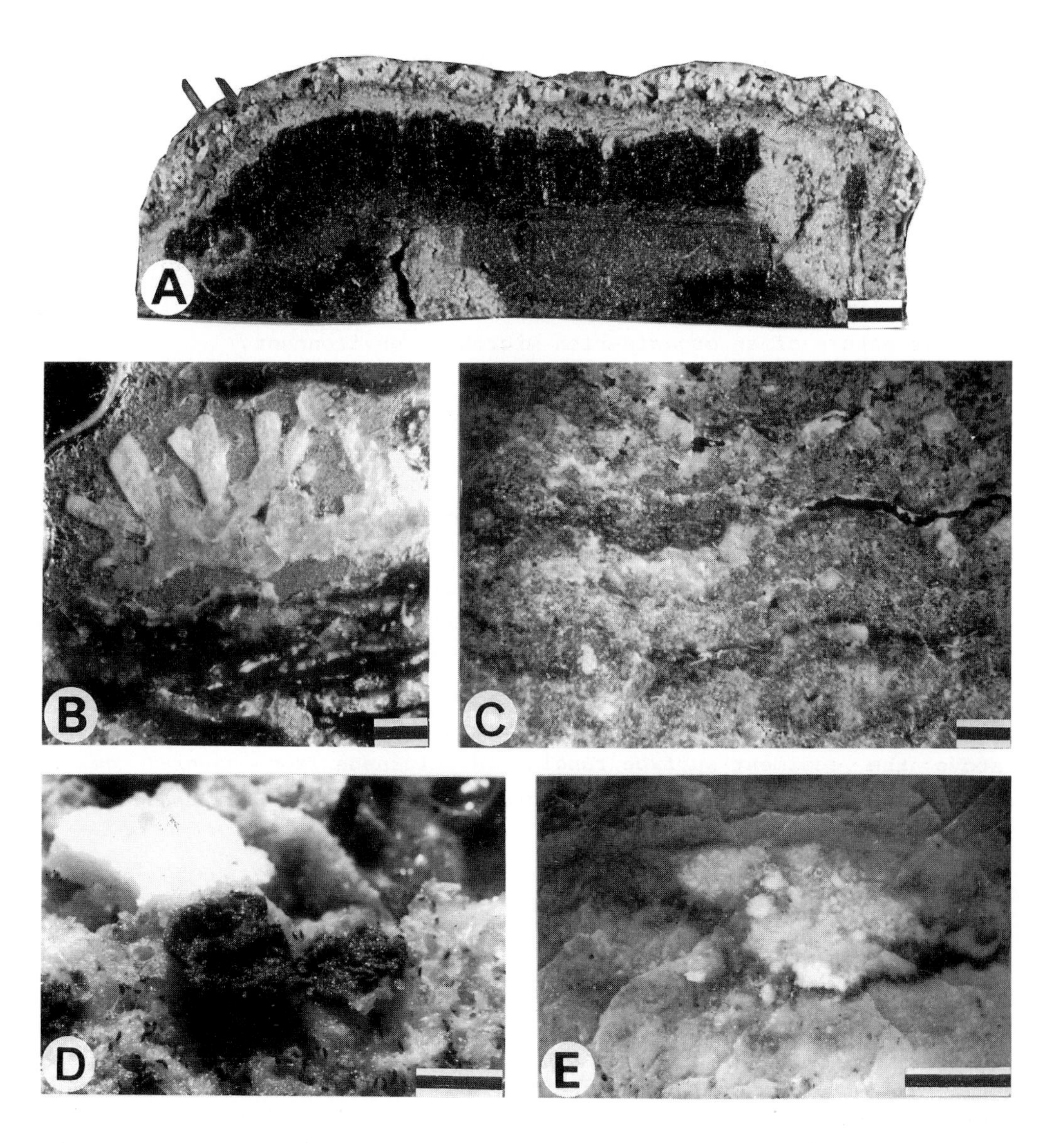

Fig. 11. Documentation of facies type 4: Biolaminated sulfate.

A) Raisin-embedded microbial mat from the lagoonary rim, laterally distorted by drying cracks. The dark-reduced zone contains carbonates, the light oxygenated topmat is interspersed with elongated gypsum crystals. Note oxic/anoxic boundary following the laterally distorted mat surface. Left: Saltbeetle burrow with oxygenated halo. Scale is 1 cm.

B) Gypsum crystals growing in a microbial mat expand the space between dark L_h-laminae. Scale is 1 mm.

C) Faintly biolaminated sulfate deposits. Scale is 1 mm.

D) Gypsum deposits, protruding above the sediment surface. Pupae of the brine fly *Ephydra* sp. are attached under the crust. Scale is 2 cm.

E) Intrasedimentary gypsum nodule, coated by a monolayered microbial mat. Scale is 1 mm. Figs. B, C, E from thin sections.

cyanobacteria (mainly *Gloeothece*, *Synechocystis* and *Synechococcus*). Each lamina type is vertically repeated several times. Up to 10 living strata thus form which harbor a multitude of oxyphotobacteria (cyanobacteria), anoxyphotobacteria (e. g. *Thiocapsa*, *Chromatium*, *Chloroflexus*), sulfate- and sulfur-reducing chemoorganotrophic bacteria. Various products of the metabolic activity of these organisms contemporaneously penetrate the vertically structured living system (KRUMBEIN et al., 1979). Ranging in thickness up to 2 cm, it reflects the multispecies nature of an organic-rich microbial environment.

4. Biolaminated sulfate (Fig. 11)

Gypsum precipitation takes place at both the amphibious bar- and center-oriented rims of the lagoon. As long as bacterial sulfate reduction predominates precipitation is restricted to oxygenated topmats (Fig. 11A). Massive gypsum banks first accumulate where the biological activity decreases due to increasing salinity. This occurs at the center-oriented rim of the lagoon (Fig. 7). Elongated gypsum crystals, sometimes twin-shaped, penetrate and fragment the interlaminated microbial mats (Fig. 11B). Faintly biolaminated sulfate deposits protrude above the sediment surface ranging in thickness from several cm to several dm (Figs 11C and 11D). Microbe-coated gypsum nodules also form (Fig. 11E).

Community structures. Unicellular cyanobacteria (mainly *Synechocystis*) dominate the microbial communities. A few filamentous forms, mainly *Spirulina*, some halobacteria and at the sulfate-sediment interface anoxybacteria, mainly *Thiocapsa* and *Chromatium*, are also present.

This facies is the record of the gradual rise of the terrane towards the center and correlates also to the further increase of salinity. Close to the lagoonary rim, salinity ranges between 220 and 300 $^{o}/oo$ in winter and between 280 and 320 $^{o}/oo$ in summer. A maximum of 360 $^{o}/oo$ is maintained in bedouin boreholes which lie just beyond the central plateau.

1. 6. 4. Facies type-related biogeochemistry

This section deals with local records of standing crops and physicochemical profiles (Eh and pH) in sediments. Measurements were made within the facies types described above: (1) the sandlobes (facies type 1), (2) the saline mud flats (facies type 2), (3) the lagoonary basin (facies type 3) and (4) the sulfate flats (facies type 4). An informal overview is given in Fig. 12.

Standing crops. Biomass studies in the Gavish Sabkha mats have been carried out by HOLTKAMP (1985). These studies reveal the relatively large standing crops (expressed in µg chl a per cc) in endobenthic *Microcoleus* mats which form the L_h-laminae in both monolayered mats (e. g. within the sandlobes; Fig. 12B section 1) and multilayered mats (e. g. within the lagoonary basin; Fig. 12B section 3). Gelatinous mats which form the L_v-laminae are usually lower in chlorophyll content (Fig. 12B sections 2 and 3). This can be attributed to the different compaction of photosynthetically active organisms within the laminae. The *Microcoleus* mat forms a heavily condensed network of sheath-encased filamentous organisms (Fig. 8D). Thus actively photosynthesizing organisms are relatively more abundant than in the gelatinous mats characteristic of a high slime to cell ratio. The slime contributes to a swelling of the biolaminite, hence the number of photosynthetic cells decreases in relation to the total bulk volume of the L_v-lamina.

Studies carried out by BAULD (1984) at Shark Bay and Spencer Gulf (Australia) demonstrate similarily that for any one mat type local variations substantially affect standing crops. Water availability plays the most decisive role, as can be seen in Gavish Sabkha, where ponding of seepage waters in the winter and desiccation in the summer take place. During the winter period the chlorophyll a content of *Microcoleus* surface mats averages ten times that of the summer mat where desiccation takes place.

The data of HOLTKAMP (1985) reveal high concentrations of pheophytin a (the degradational product of chlorophyll a) in almost all mat types. Concentrations of pheophytin a up to 5 x greater than those of chlorophyll a were found in L_v mats. This indicates that the growth of photosynthetically active organisms is accompanied by rigid degradation processes. A chlorophyll a / pheophytin a ratio (C/P ratio) smaller than 1 suggests for several microbial mats of the Gavish Sabkha that degradation of chlorophyll a takes over already simultaneously with the photosynthetic activity in the same surface layer (Fig. 12B sections 2, 3 and 4; values calculated by HOLTKAMP).

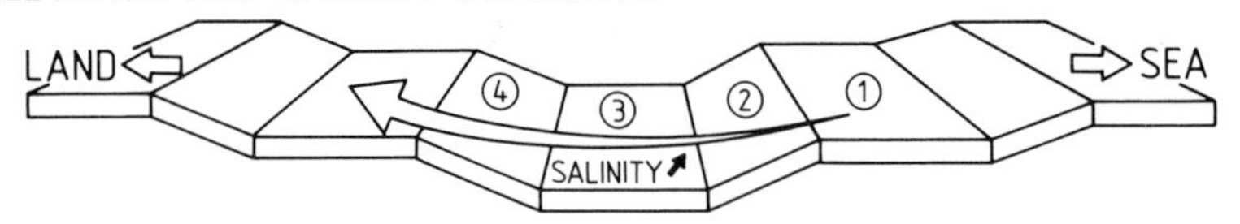

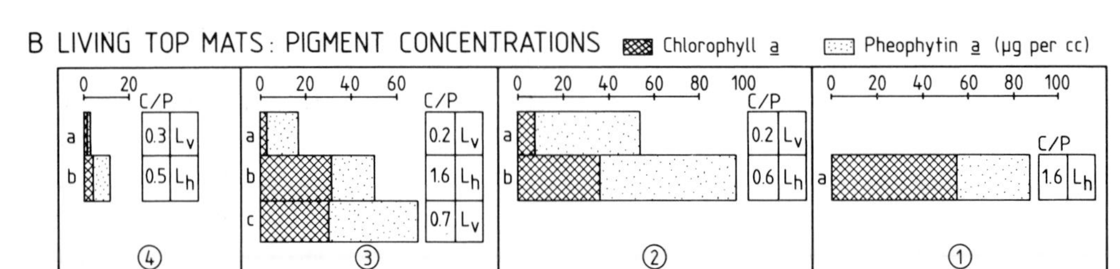

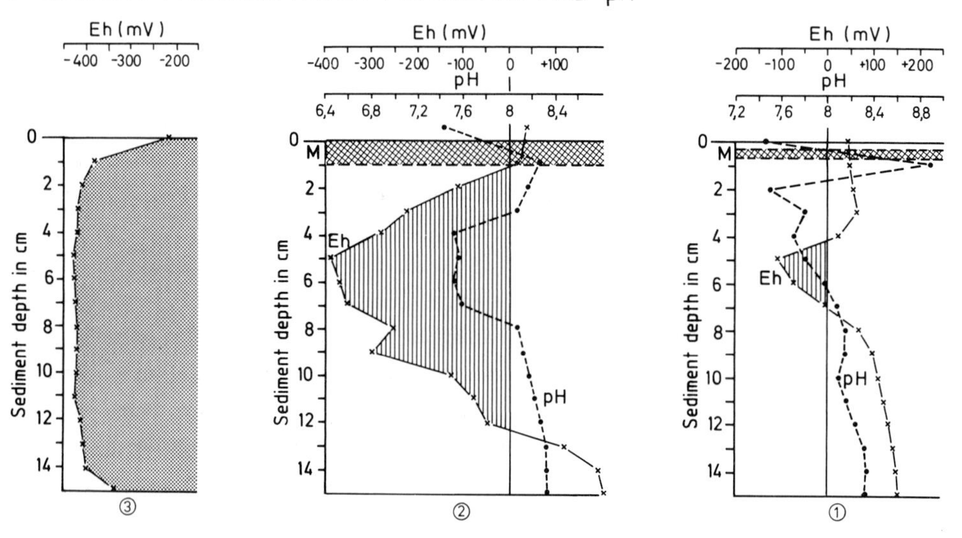

Fig. 12. Effects of terrane inclination and salinity increase on standing crops (data after HOLTKAMP, 1985) and physicochemistry.

A. Schematic presentation to show inclination and salinity increase.

B: Chlorophyll a, pheophytin a (degradational product of chlorophyll a) and C/P ratios in surface mats. The tendency of microbial communities to build multilayered mats increases with the inclination of the terrane and the salinity increase. Sandlobes (1): Mats are monolayered and of the L_h-type. Saline mud flats (2), lagoon (3) and gypsum flats (4): Mats consist of two or three laminae. Lowest pigment concentrations within the gypsum flats (4) and L_v-top mats (see 2a, 3a). C/P ratios smaller one indicate that pheophytin a takes over. Highest C/P ratios occur in L_h mats (see 1, 3b).

C: Eh- and pH-profiles in sediments (M = surface mat). Reduced horizons are dashed or shaded to show their increasing vertical extension.

Reduction-oxidation potentials and pH. Physicochemical properties in terms of Eh and pH are summarized in Fig. 12C which relates to the areas of standing crops described above.

Within the siliciclastics of the sandlobes (areas of facies type 1) Eh-values were found mainly in the positive sector. Small peaks reaching into the negative sector indicate buried mats where bacterial sulfate reduction prevails (e. g. from -4 to -6 cm below surface in Fig. 12C section 1). The pH in photosynthetically active topmats frequently reaches values approaching 9 and falls back in the seawater-soaked sediments below to slightly alkaline marine conditions.

On the mud flats (areas of facies type 2) only positive Eh-values are found in the surface mats. The sediments below are strongly reduced up to a depth of -12 cm under the surface. Eh-values of up to -400 mV (Fig. 12C section 2) were measured.

The sediments of the lagoon (localities of facies type 3) are likewise strongly reduced right up to the surface (Fig. 12C section 3). pH-Values were between 7.4 and 8.2. However, extremely low values of pH 6.0 to 6.5 were observed by E. HOLTKAMP in the black anaerobic sediments (Holtkamp, 1985). The more continuously the sediment surface is saturated or covered by water, the higher the biological activity of the photosynthetic microorganisms is. These organisms form the organically rich sediments which are consequently colonized by chemoorganotrophic bacteria which decompose the dying organic material, with the help of sulfate as an electron donor. The sulfate is supplied by the seepage seawater and correlates in its concentration well with the topographic moisture gradient (see Fig. 3). The increase in the concentration of physiologically derived reductants within sediments of the strongly hypersaline lagoon shows that salinity is a secondary factor while moisture is of extraordinary importance for the microbial primary producers.

1. 6. 5. Products of early diagenetic processes

The definition of early diagenesis refers to changes within sediments after burial (BERNER, 1980). Besides burial by sheet flood deposits, self-burial by vertical successions of microbial mats is important. This is achieved by the active positioning of microbial populations relative to their environmental requirements, e. g. with respect to the light field. The rapid gliding motility (up to 3 cm/h) of *Oscillatoria* sp. a common filamentous species in the Gavish Sabkha, and its penchant for spreading out under optimal conditions of light and temperature allows this species in particular to override other cyanobacterial topmats (CASTENHOLZ, 1969). Also the filamentous cyanobacterium *Microcoleus chthonoplastes* is able to perform environmental stimulus-induced movement and accumulates at sediment surfaces under reduced light conditions (HOLTKAMP 1985). This species usually moves back and forth in its bundled sheath and builds up horizontal bundles in a complex interactive way (L_h-lamina type). In crisis conditions trichomes (and hormogonia; short filaments of undifferentiated cells) move faster and further away from the growth site.

The organic matter left behind exhibits great diagenetic variety This is evident even in the topmost layers. The following features described below can be related to early diagenetic processes:

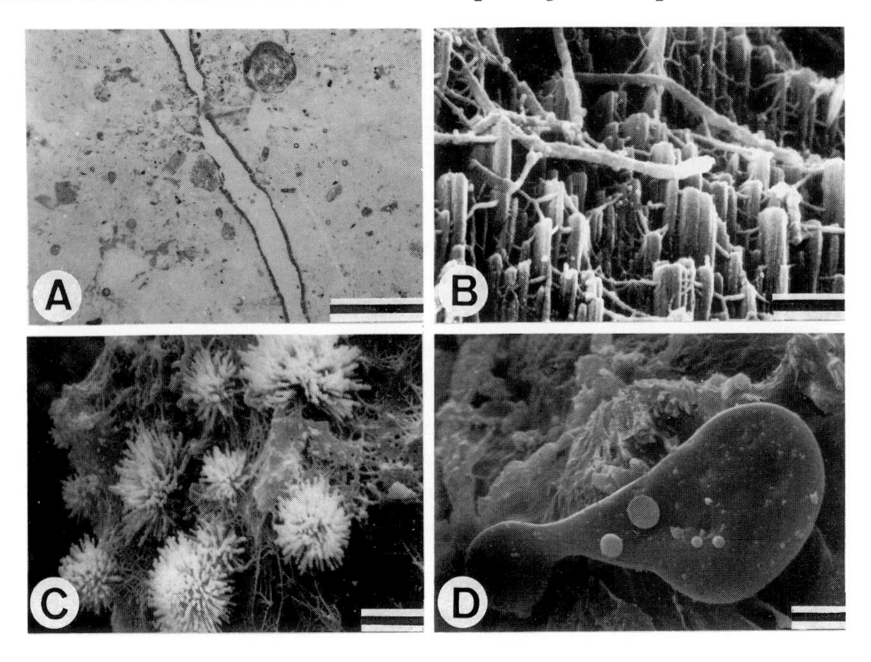

Fig. 13. Patterns of dehydration and dissolution in microbial mats.
A) Synaeresis-crack running through an extended hydroplastic L_v-lamina. The dark coating may be caused by subsequent microbial colonization. Thin section from sediment core. Scale is 1 mm.
B) Partial dissolution of gypsum overgrown by a microbial mat. Filaments of chemotrophic bacteria and slime threads coat the crystals. Scale is 2 µm (B to D: SEM-photography).
C) Pyrite crystals cover bacterial slime. Scale is 3 µm.
D) Partial dissolution of a halite chip. Scale is 5 µm.

(1) Synaeresis cracks (Fig. 13A)

Synaeresis cracks are generally thought to be characteristic of gel-like material (WHITE, 1961; BURST, 1965) and have been attributed to spontaneous dehydration even in subaqueous sediments (PETTIJOHN, 1975). In the Gavish Sabkha sediments synaeresis cracks occur within the extended hydroplastic layers composed of extracellular polysaccharides (Fig. 13A). Even the muculous clouds responsible for the viscous appearance of the surface water contribute to these layers due to gravitational displacement.

(2) Dissolution patterns of evaporite minerals

Several dissolution patterns are visible on the surface of embedded evaporites. Dissolution of gypsum in buried mats (Fig. 13B) may be more abundant where bacterial sulfate reduction occurs. Complete coating of gypsum with bacterial slime is evident. Cobwebs of extracellular mucus are commonly covered with various pyrite rosettes as shown in Fig. 13C. In contrast, the shape of the halite chip in Fig. 13D indicates partial dissolution resulting from the diffusion of salt into porewater fluids of lower concentrations.

(3) Decay centers and formation of authigenic minerals

The term authigenic mineral refers to mineral species which grow within a sediment after burial (BERNER, 1980). By contrast to the in situ formation of minerals, allochems are formed elsewhere and are transported to the deposits in question where they accumulate. A causal factor in the formation of authigenic carbonates occuring in buried Gavish Sabkha mats appears to be the bacterial decomposition of organic material left behind after death (BERNER, 1980).

The sheaths of *Microcoleus chthonoplastes* which form the L_h-laminae are particularly recalcitrant and survive decomposition. BOON et al. (1985) have found that the sheath material consists of biopolymers which are evidently resistant to microbial attack. The sheath remains after the cyanobacterial filaments have disappeared. Thus L_h-laminae composed of sheaths are often retained even in deeper sediment layers (COHEN, 1984). Elongated arrangements of micrite crystals point to the former presence of filamentous organisms (Figs. 14A to D). Capsules formed by coccoid unicells also possess a high degree of recalcitrance

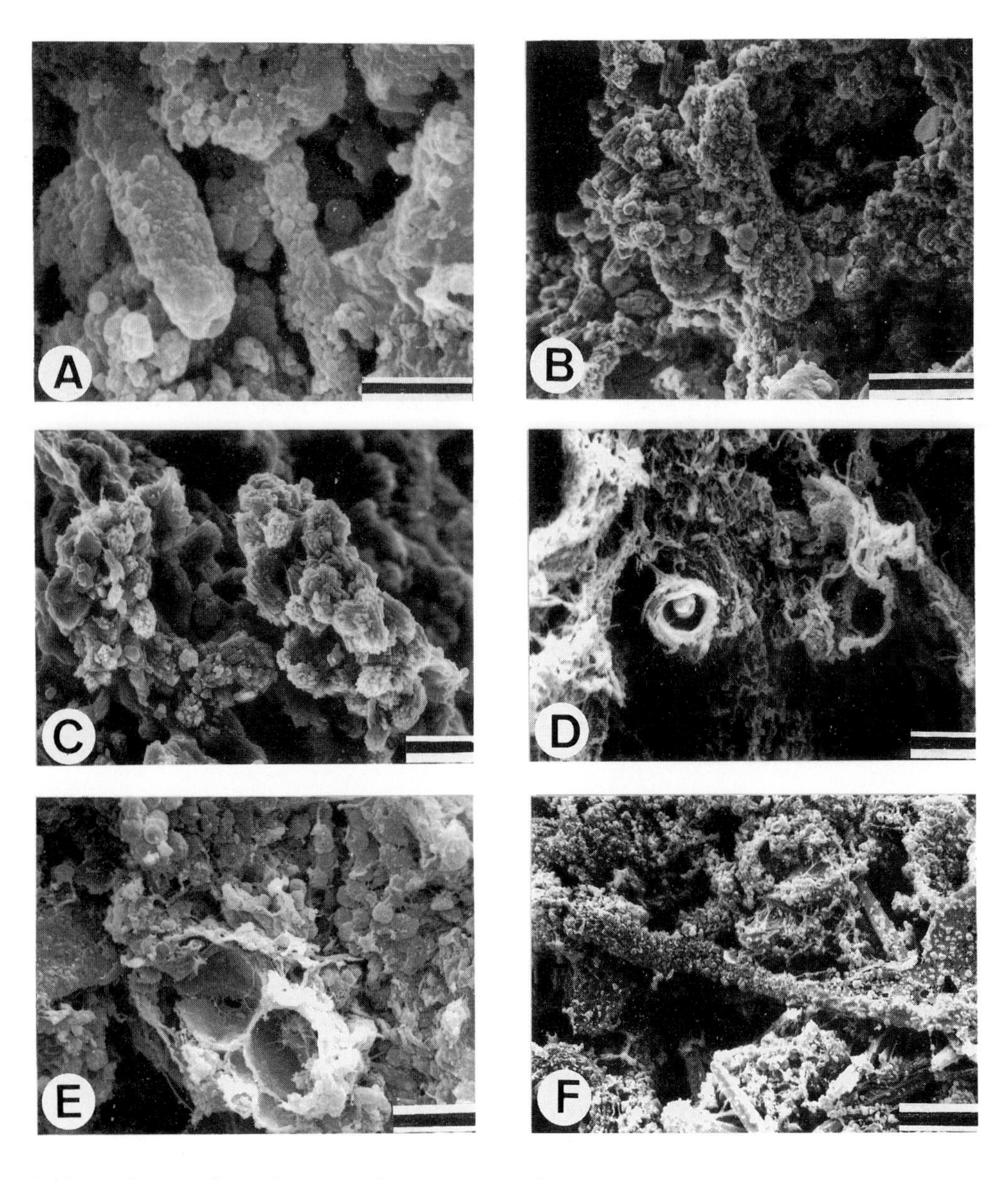

while the enclosed protoplasts are subject to more rapid decay. The resulting honey-comb pattern of hollow spheres is a characteristic feature of decaying mats (Figs. 14E and F).

A coccoid colony without capsules (Fig. 15A) may give rise to a micrite-surrounded hollow sphere after decomposition. Coccoid unicells showing early stages of micritization are illustrated in Fig. 15B.

Fig. 14. SEM-photographs showing localized decay and carbonate precipitation in buried mats at 60 to 80 cm sediment depths (A, B, C, F) and about 1 cm below surface (D, E).
A) Granular mg-calcite precipitate around decaying trichomes of cyanobacteria. At the lower end of the left tube is its empty state visible. Scale is 5 µm.
B) Continued precipitation of microcrystalline mg-calcite replaces gradually the filamentous form. Scale is 25 µm.
C) Only the elongated arrangement of micrite crystals points to the former filament. Scale is 10 µm.
D) Polysaccharid capsules are more resistant to bacterial attack than protoplasts: Degrading cell encased by the extracellular capsule. Scale is 10 µm.
E) Empty capsules of cyanobacteria (probably Pleurocapsalean) surrounded by smaller coccoids (possibly sulfur bacteria). Empty capsules later give rise to micrite-surrounded hollow spheres. Scale is 10 µm.
F) View of the decaying mat with coated filaments, diatoms and hollow spheres. Note the slightly deformed coated sphere at top. Such spheres also form around gas bubbles abundantly occurring within the decaying mat. Bubbles colonized, coated and fixed are common (see also Fig. 26). Scale is 25 µm.

Single cells appear as completely calcified spheres (Fig. 15C). Ooids and oncoids of various shapes and sizes (Figs. 15D, E and F) are also records of carbonate precipitation within the buried mats (see next section).

1. 6. 6. In-situ formation of ooids and oncoids

Self-burial, emphasized in the previous sections, is recorded mainly within the facies type 3 which occurs in the constantly submerged lagoon. It results in the vertical succession of dark (filamentous supported L_h-) and light (unicellular and gel supported L_v-type) microbial mats. The study of thin sections shows that ooid and oncoid populations predominantly occur within the light L_v-laminae (Fig. 10). In L_h-laminae carbonate grains are rare. In sheetflood sediments and evaporites, which interfinger horizontally and vertically with the facies type 3, they are non-existent. Their close relationship with the L_v-laminae will be examined here. It is stressed that the ooids are not forming in the off-shore waters nor are they found in the layers rich with allochems (compounds that have been transported into the environment).

The term ooid is applied to a rounded biogenic grain with concentric and continuous laminations; it is less than 2 mm in diameter (Table 2).

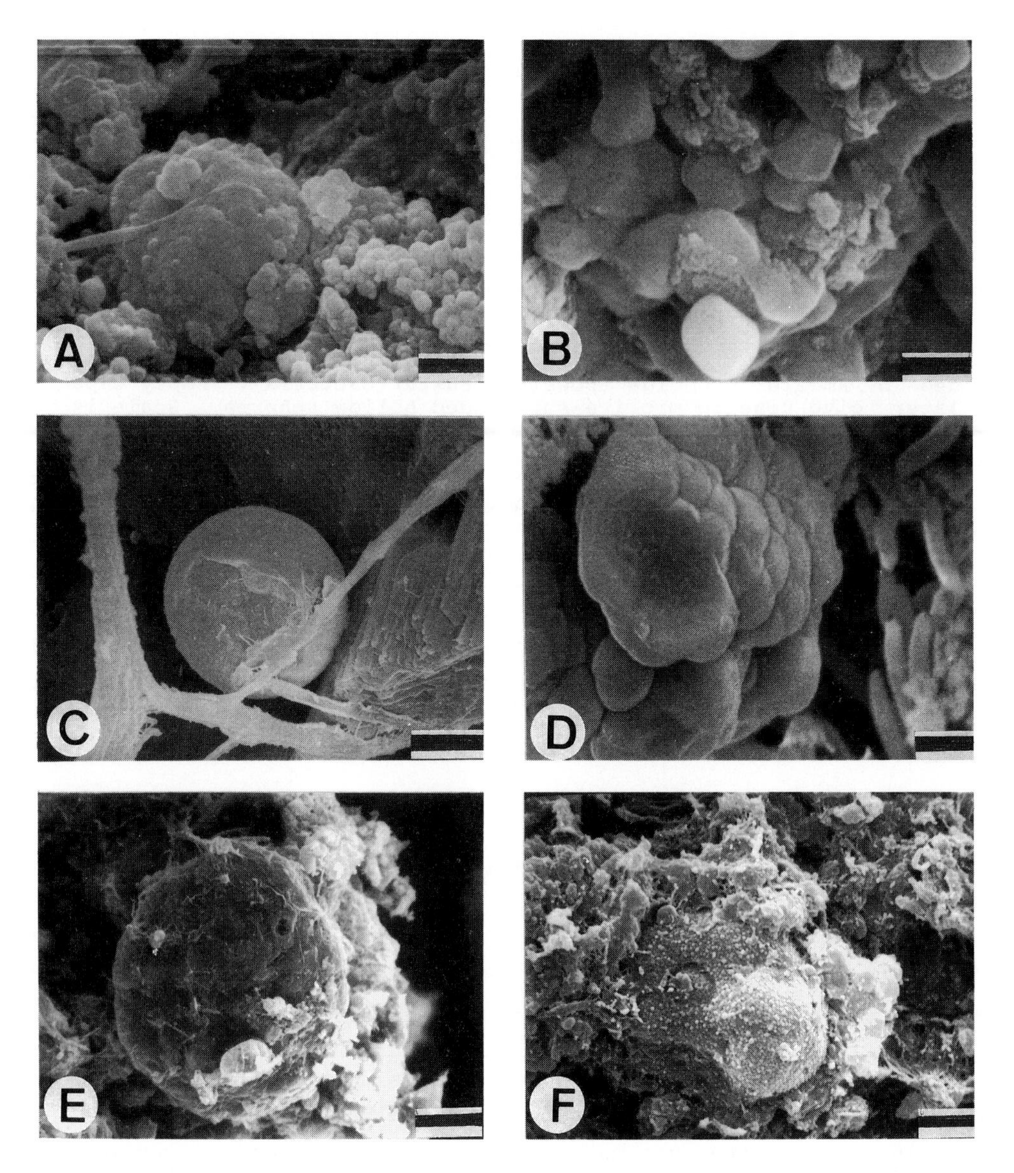

Fig. 15. SEM-photographs showing cells and rounded cell clusters in mats which may give rise to nucleation centers of ooids, compound ooids and oncoids.

A) Colony of coccoid unicells partially agglutinated showing initial steps of calcite coatings. Scale is 2 µm.
B) Microtexture of a colony of coccoid unicells during an early stage of lithification. Scale is 2 µm.
C) Completely calcified single cell. Scale is 5 µm.
D) Sub-rounded shape of lithifying cell colony. Rod-shaped organisms at right may be chemoorganotrophic bacteria. Scale is 2 µm.
E) Repetition of microbial coating of a completely round carbonate grain. Scale is 10 µm.
F) Sub-rounded grain embedded in a slime matrix and colonized by small unicellular bacteria. Scale is 3 µm.

Where the laminations are discontinuous and irregular and the grains are less rounded, the term oncoid will be applied (HEIM, 1916, see also DAHANAYAKE & KRUMBEIN, 1986).

In-situ formation of ooids in particular may be striking but can be correlated to conditions already mentioned in the previous section. These conditions are summarized as follows:

(1) Hydroplastic L_v-laminae provide the space:

In the vertical sequence much more extensive L_v-laminae repeatedly override the thinner L_h-laminae, which may be flat, wavy or "eye"-shaped (Fig. 8). The L_v-laminae are characterized by large quantities of gel. The mucilage tends to become irregularly disintegrated and partially diluted by interstitial seawater. Small-scale pockets and vugs develop due to irregularities in the slime matrix and the underlying L_h-laminae.

(2) Nucleation centers:

On a microscopic scale, the widely spaced L_v-laminae contain a variety of possible nuclei for ooid and oncoid synthesis: trapped gas bubbles, intraclasts, degrading cell clusters. The bubbles often come from L_h-type mats below. These generally have a higher primary production rate than the L_v-laminae above. The filamentous organisms dominant in L_h-laminae are able to photosynthesise even under reduced light conditions. Metabolic products (e. g. O_2, CO_2, NH_3, CH_4, H_2S), become trapped during upward migration within the hydrophilic gels which lower their diffusion velocity by at least the factor ten (KRUMBEIN et al., 1979). The intraclasts which can be repeatedly observed within the laminae originate from floating and/or fragmentated mats. Finally, unicellular cyanobacteria, which are mainly responsible for gel production, tend to form cell clusters rather than regular arrangements within the slime-supported layers and present ideal nucleation centers after death (Fig. 15).

(3) Nucleation and continued growth of ooids and oncoids:

Bubbles, intraclasts, filaments, single cells, cell clusters and organic fragments, often on submicroscopic scale, may serve as nuclei for $CaCO_3$ precipitation. Initial $CaCO_3$ precipitation may be due to the localized rise in carbonate alkalinity as a result of ammonia and CO_2 supersaturation due to bacterial decomposition of proteins. The reactions given by BERNER (1980) are as follows:

organic C $\rightarrow CO_2$
organic N $\rightarrow NH_3$
$2NH_3 + CO_2 + H_2O \rightarrow 2\ NH_4^+ + CO_3^=$

Dissolved sulfate which is supplied in high concentrations by the interstitial brine is constantly reduced to sulfide by sulfate-reducing bacteria. This additionally generates carbonate alkalinity:

$2CH_2O + SO_4^= \rightarrow H_2S + 2HCO_3^=$

Thus, bacterial attack on protoplasm material generates highly reactive zones within the host material of extracellular gel which itself is more recalcitrant against decay. These highly reactive centers are commonly of microscopic or even submicroscopic scale so that the rather rough measurements of general sediment parameters shown in section 1. 6. 4. do not permit their recognition. In situ microelectrode measurements can solve this problem. The effect of localized centers of alkalinity is a rise in the level of $CaCO_3$ oversaturation. As a result, $CaCO_3$ precipitates around the decomposing cells, fractalized protoplasm material and other centers of concentrated ions (e. g. bubbles filled initially with CO_2 or CH_4). The effect of encapsulation may be analogous to the formation of perls, bezoar (kidney) stones and calculi found in tissues of macroorganisms as a reaction to disturbances.

The continued growth of concentric layers may be due to external factors (overall saturation of the concentrated brine with carbonate and sulfate ions), but it seems worthwile to emphasize the conspicuous habit of microorganisms living in the surrounding matrix of colonizing any given "hard ground" (see for example the coating of extraclasts in Fig. 8). Thus, the continued growth of calcareous layers around initially calcified soft-bodied material appears to be achieved by subsequent colonization, which again accelerates the reaction chain after death: decomposition, production of excess free energy and precipitation. This process may bring about the successive laminations of the ooids and oncoids.

Finally, the question of the shape of pre-existing soft-bodied nuclei appears to be important. Unicellular organisms (Fig. 15C), bubbles (Fig. 14F shows for example the coating of a bubble by mat inhabitants), and smaller lithified cell aggregates (Figs. 15A, B) may bring about the concentrically continuous ooid type (Fig. 15E), while the oncoid type is brought about by less rounded nuclei (Figs. 15D and F).

All the facts mentioned above as well as other physiological and biogeochemical studies of microbial mat systems (KRUMBEIN, 1978, 1983; COHEN et al., 1984) allow us to infer the possible syngenetic growth of ooids and oncoids within stromatolitic sequences. It should, however, not be construed that all cases of ooid and oncoid formations are covered by these models.

1. 6. 7. Microbially modified surface structures

Subaerially exposed and salt-encrusted surfaces bearing a number of physical structures such as desiccation cracks and crinkle marks are included here. The aim is to demonstrate how physical structures can be modified by the presence of microbes.

(1) Tepee- and petee structures

The central elevated part of the Gavish Sabkha is characterized by desiccation cracks which form polygons. Tepee structures (sensu ADAM & FRENZEL, 1950) are common here (Fig. 16A). These result from crystallization pressure in precipitating salts, which causes a lateral expansion of surface sediments (SHINN, 1969; WARREN, 1982). Tepees and polygons in the central part of the sabkha were found without any sign of microbial life.

On higher-lying parts of the seaward-directed slope, however, microbes develop mats episodically after sheetfloods or when groundwater levels rise. For a period of several days, terrigenous sediment particles become glued together by microbial slime and precipitating salts. During subsequent periods, the sediments dry out completely and crystallization pressure expands them as on the surface of the central part. According to the microbial influence, modified tepees result which show somewhat crinkled and convolutional patterns. In analogy to

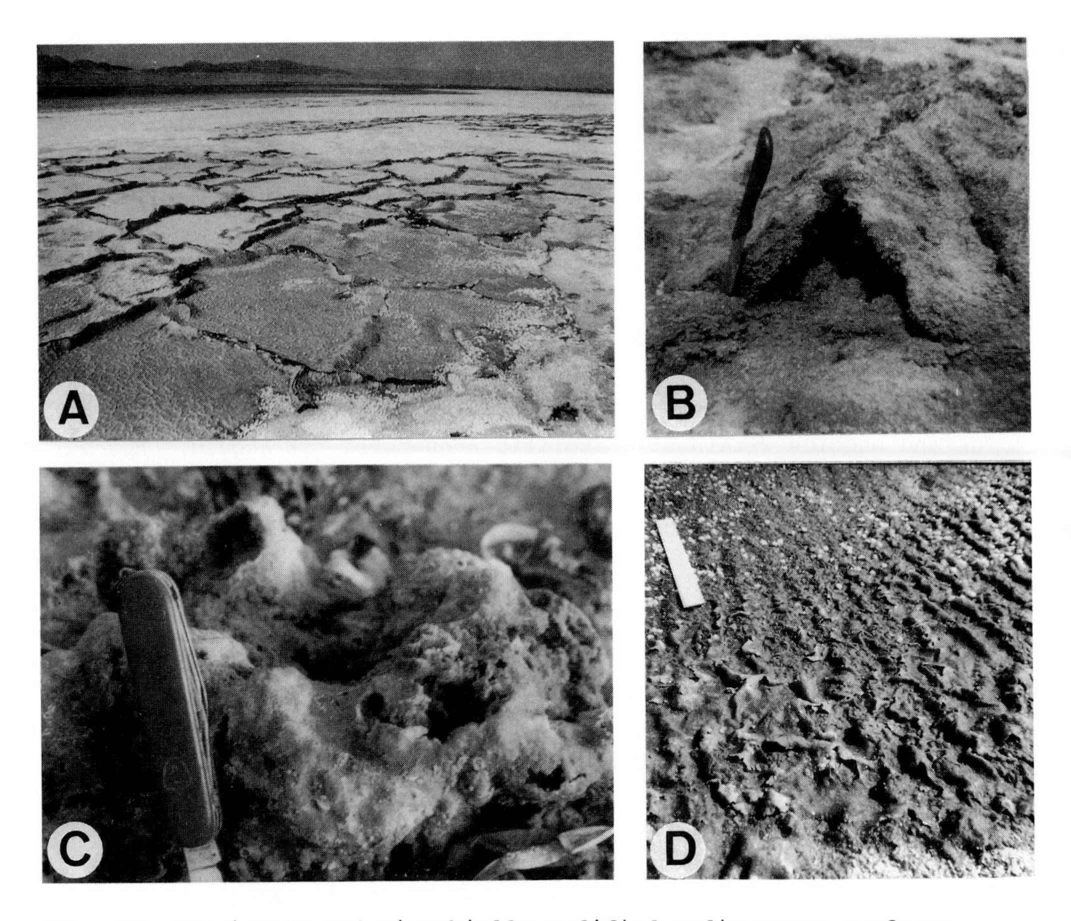

Fig. 16. Physically and microbially modified sedimentary surfaces.
A) Desiccation polygons and tepee structures, typical for sabkhas, are physical in origin. Elevated center of the Gavish Sabkha.
B) Close-up of a tepee structure.
C) Crystallization pressure and lateral expansion of surface sediments take also place where microbial slime binds surface sediments. The result is a convolutional pattern (modified tepees or "petees").
D) Sediments glued together by microbial slime form slump structures like a table cloth when moving downslope. Cracking takes place along the crests after drying. Finally "meteor-paper" results (see text).

tepees which originate under merely physical conditions, we call the microbially modified structures "petees" (Fig. 16B).

(2) Pseudo-ripples and meteor-paper

On gently sloped, periodically moistened surfaces, slump structures occur which result from a restricted downslope movement of sediments, glued together by microbial slime and precipitating salt (mainly gypsum). The structures formed include small microfolds perpendicular to the direction of the movement (Fig. 16C). After drying, cracking starts along the crests of the hollow pseudo-ripples, and finally paper-like chips develop which can be blown off by the wind.

The product was first described in 1686 and was initially identified as meteor-paper (organic, paper-like meteorite; i. e. fallen from the sky). People saved it so that about 150 years later, C. G. EHRENBERG (1839) was able to study it again and found that it was a dried chip of microbially glued sediments blown off by the wind (see also KRUMBEIN, 1986b).

1. 7. Faunal influence on the biolaminated deposits

The Late Precambrian decline of stromatolites coincides in time with the early Metazoan rise (AWRAMIK 1971, 1981). Ordovician decline especially of carbonate shelf stromatolites coincides in time with the radiation of bryozoans, rugose and tabulate corals and articulate brachiopods (KNOLL & AWRAMIK, 1983; PRATT, 1982). It has been suggested that the first step in the "Götterdämmerung" of stromatolites was induced by increasing bioturbation and grazing activity of benthic invertebrates (for controversal discussion of the Late Precambrian decline see MONTY, 1973; KNOLL, 1985a, b, c) and the second by competition for space. Whether there was a correlation or not, the fact is that Phanerozoic stromatolites and Recent potential stromatolites are rarely if ever found in coexistence within the known diversity of shallow marine and intertidal burrowing faunal habitats. Microbial communities usually occur in marginal to moderate, fauna-populated environments. Zones where they occur are often characterized by schizohaline, hypersaline or hyperthermal conditions. These are desiccation-endangered and thus maintain physiological barriers which are hard to overcome by a marine fauna. The Gavish Sabkha presents no exception, particularly as the migration of faunal elements from the adjacent Gulf is even more effectively restricted by both the physiological barrier of hypersalinity and by the geomorphic barrier of bar-closing.

Nonetheless, a variety of faunal elements is present in the Gavish Sabkha. It is the purpose of this section to elaborate on their facies-indicative role.

1. 7. 1. Species composition and distribution

Sampling provided 22 macrobenthic and 10 meiobenthic species (Table 4). About 80 % of all species are of terrestrial origin, 50 % (17 species) being insects.

Like the microbial mats, the benthic invertebrates of the Gavish Sabkha are restricted to the zones periodically or constantly supplied by water. While salinity is rarely a restrictive factor in the spreading of microorganisms, it plays a decisive role in the distribution of fauna. According to the topographic moisture and salinity gradients three major species assemblages can be recognized (Fig. 17): (1) terrestrial arthropods (insects, spiders) which colonize the air-exposed wetlands around the aquatic zones (Unit I); (2) aquatic insects, marine gastropods, ostracods, copepods, nematodes, turbellarians and rotifers which prefer seawater springs and small impondments at the lagoon's margins (Unit II); (3) aquatic insects which spread all over the aquatic and water-saturated areas (Unit III). Unit II is restricted by a salinity increase above 130 $^{o}/oo$ and thus has been defined hyperstenohaline (GERDES et al., 1985d) while Unit III is defined hypereuryhaline because its members manage to populate the entire salinity gradient (Fig. 17). The remarkable evenness in spatial distribution of the fauna between seasons documents well the stability of biotopic conditions due to the constant seawater influx (compare Fig. 17A and B).

1. 7. 2. Trophic relations

Gut content analyses have shown that the majority of species living in the Gavish Sabkha belongs to the category of primary consumers. This category can be further split into (1) types which prefer a diet of diatoms and unicellular cyanobacteria, (2) types which favor filamentous cyanobacteria and (3) types which feed on the decaying organic material of microbial mats. The largest part of the primary consumers feeds upon diatoms and unicellular cyanobacteria which are available in abundance in the Gavish Sabkha. The heavily ensheathed bundles of *Microcoleus* are seemingly less nutritious. Only one species favors this diet (Table 5). Secondary consumers are pseudoscorpions and spiders, the tiger beetles (both larvae and imagines) and larvae of hydrophilide beetles (*Anacaena* sp.). Capturing prey, pseudoscorpions and *Anacaena* larvae have been commonly observed in dwelling burrows of staphilinid

TABLE 4. Benthic invertebrates observed in the Gavish Sabkha

MAJOR TAXA	SPECIES	STAGES	HABITAT
1. INSECTA COLEOPTERA			
Staphilinidae	*Bledius capra*	L,P,A*)	exposed wet
	Bledius angustus	L,P,A	exposed wet
Cicindelidae	*Lophyridia aulica*	L,P,A	exposed wet
Georyssidae	*Georyssus* sp.	A	exposed wet
Ptiliidae	*Actidium* sp.	A	exposed wet
Dytiscidae	*Deronectes* sp.	L,P,A	submersed
Hydrophilidae	*Enochrus* sp.	L,P,A	submersed
	Anacaena sp.	L,P,A	submersed, exp.wet
Hydraenidae	*Ochthebius* c.f. *auratus*	L,P,A	submersed
2. INSECTA DIPTERA			
Ephydridae	*Ephydra macellaria*	L,P,A	submersed, exp.wet
	Hecamede grisescens	L,P,A	submersed
Muscidae	*Musca crassirostris*	L	exposed wet
	Orthellia caesarion	L	exposed wet
Tabanidae	*Atylotus agrestis*	L,A	submersed
Ceratopogonidae	*Bezzia* sp.	L,P,A	submersed
3. INSECTA HETEROPTERA			
	not determined	J	exposed wet
4. ARACHNIDA			
Pseudoscorpiones	*Halominniza aegyptiaca*	J,A	exposed wet
Clubionidae	not determined	J,A	exposed wet
5. CRUSTACEA			
ISOPODA	*Halophiloscia* sp.	J,A	exposed wet
OSTRACODA	*Cyprideis torosa*	J,A	submersed
	Paracyprideinae sp.	J,A	submersed
COPEPODA	*Robertsonia salsa*	J,A	submersed
	Nitocra sp.	J,A	submersed
6. ANNELIDA			
CLITELLATA	Enchytraeidae, n. d.	J	exp.wet
POLYCHAETA	Capitellidae, n.d.	J	exp.wet
7. MOLLUSCA GASTROPODA			
Potamiidae	*Pirenella conica*	J,A	submersed, exp.wet
8. NEMATODES	*Enoplus communis*		submersed, exp.wet
	Oncholaimus fuscus		submersed, exp.wet
	Adoncholaimus oxyris		submersed, exp.wet
9. ROTATORIA	*Brachionus plicatilis*		submersed
10. TURBELLARIA	*Macrostomum* sp.		submersed
11. PROTOZOA			
CILIATA	not determined		submersed, exp.wet

*) observed stages: L=Larvae, P=Pupae, A=Adults, J=Juvenile

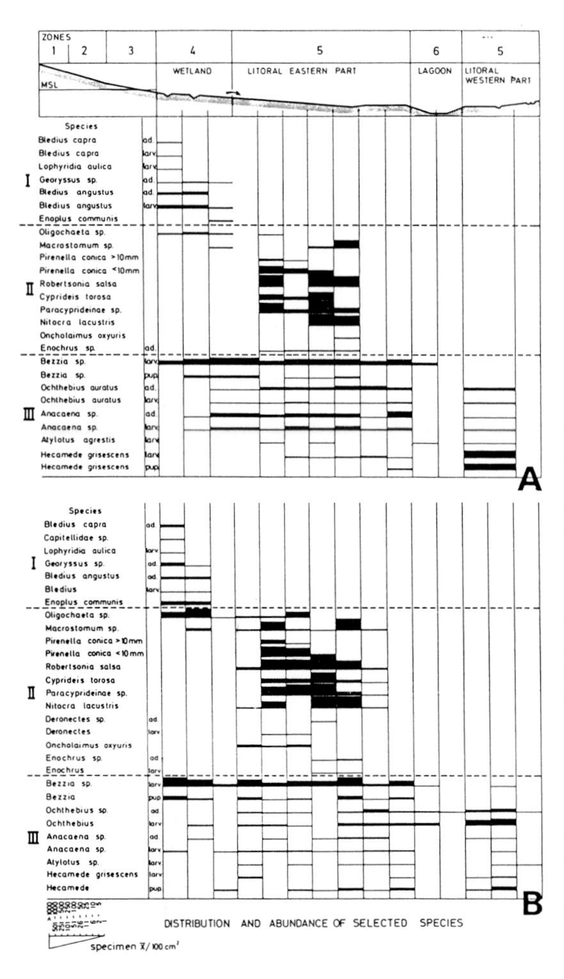

Fig. 17. Distribution and abundance of benthic invertebrates (selected species) along horizontal gradients of water supply and salinity. A) summer B) winter situation. Unit I: Wetland fauna, restricted to exposed flats (e. g. the sandlobes); Unit II: Aquatic fauna, salinity-restricted (hyperstenohaline), members of this group are restricted to mud flats and small impondments of the eastern (coastal bar directed) rim of the lagoon; Unit III: Aquatic fauna, non-restricted by salinity (hypereuryhaline). Members of the latter group (only insects) manage to populate the entire salinity gradient. (Modified after GERDES et al., 1985).

beetles. Dytiscid beetles which belong to the hyperstenohaline unit II are predators as well.

1. 7. 3. Systematic ichnology

Dwelling, crawling and feeding traces account for ichnological patterns in the Gavish Sabkha.

A. Dwelling traces:

1. Traces of the salt beetle *Bledius capra* (Staphilinidae)

Description: A diagonally oriented "bottle neck" starts from the surface and converts into a vertically oriented burrow (Fig. 18A, B). Length of bottle neck 2 to 3 cm, of complete burrow 10 to 12 cm. Diameter: bottle neck about 5 mm, vertical burrow about 1 cm. At the conversion there is a widening where microbial lumps are stored (Fig. 18C). The lower end of the vertical burrow merges into one or more appendices where the beetle stores feces. Female beetles interrupt the walls of the vertically oriented burrow to create breeding chambers (Fig. 18B). During breeding each chamber contains one egg which is vertically oriented and stands on a sockle of sediment grains.

Habitats: Gully bottoms and sandlobes, air-exposed microbial mats.

Trophic relations: Grazer.

Remarks: The beetles rip away at the tough substrates of air-exposed surface mats and carry the food into the store chambers where they use mandibles (Fig. 18D) and heads to press the lumps against the wall. Positioning the mat material close to the surface enhances the chances of photosynthesis. Sheetflood sediments covering multilayered microbial mats are well documented by burrow fillings (Fig. 18 E).

2. Traces of the salt beetle *Bledius angustus* (Staphilinidae)

Description: Traces consist of a combined system of (i) vertically oriented endobenthic burrows, (ii) horizontally oriented feeding traces, parallel to bedding planes (iii) vertically oriented epibenthic "chimneys" (Fig. 19A). Length of endobenthic burrows: up to 5 cm, diameter up to 5 mm. Length of epibenthic chimneys: up to 3 cm, diameter up to 1 cm (incl. wall). Chimneys are only present when gypsum precipitates on the surface and glues the sand grains together (Fig. 19A). Length of feeding traces about 5 cm, diameter 5 to 10 mm, transverse U-shaped.

Habitats: Gully bottoms and sandlobes consisting of loose or salt-encrusted sand.

TABLE 5. Trophic relations of benthic invertebrates present in the Gavish Sabkha (after studies of gut contents)

Groups	Species	Food type (preference)			
		Coccoid cyanobact. and diatoms	Filamentous cyanobact.	Detritus	Prey
Terrestrial beetles	*B. capra*	x			
	B. angustus	x			
	Georyssus sp.	x			
	Actidium sp.	x			
	L. aulica				x
Aquatic beetles	*Anacaena* sp.	x			
	(adult)	x			
	(larvae)				x
	Deronectes sp.				x
	O. auratus	x			
	Enochrus sp.	x			
Flies and mosquito larvae	*H. grisescens*	x			
	E. macellaria	x			
	Muscidae			x	
	Bezzia sp. (larvae)	x			
Spiders, Pseudo-scorpions	Clubionidae				x
	H. aegyptiaca				x
Isopods	*Halophiloscia* sp.			x	
Copepods	*R. salsa*	x			
	Nitocra sp.		x		
Ostracods	*C. torosa*	x			
	Paracypri-deinae	x			
Gastropods	*P. conica*	x			
Nematodes	*E. communis*	x			
	Adoncho-laimus sp.	x			
Turbel-larians		x			
Rotatorians		x			

Trophic relations: Grazer.

Remarks: Feeding structures run -2 to -5 mm below surface and furrow the endobenthic microbial mats of gully beds and sandlobes. Where evaporation crusts replace tunnel roofings feeding traces furrow the sediment surface. SCHWARZ (1936) suggests that the use of prefabricated buildings econimizes on construction effort. Evaporation crusts also serve as light-channelling systems which stimulate photokinesis of cyanobacteria which then colonize the furrows made by the beetles. Hence a farming effect evolves from an interplay of behaving beetles and reacting cyanobacteria, the latter create a dense bluegreen carpeting which can easily be grazed under the protection of salt crusts at the sediment/air interface. The beetles manage to survive when the groundwater table rises by retreating into the epibenthic chimneys, while the endobenthic burrows (retraite-burrow sensu BRO LARSEN, 1936) support survival when it falls. The whole system evokes the impression of a three-dimensional cross (Fig. 20B) which allows the beetle to access the horizontally oriented cyanobacterial "garden" from both the supersurficial and subsurficial parts of the burrow.

The presence of both *Bledius* species can be recognized by the typical piles of excavation pellets they leave behind on the mat or sediment surface respectively (Fig. 18F).

3. Traces of tiger beetle larvae (*Lophyridia aulica*, Cicindelidae)

Description: Vertically oriented burrows, length up to 30 cm, diameter up to 1.5 cm.

Habitats: Gully embankments consisting of loose, sand-sized sediments with a low moisture level.

Trophic relations: Predator.

Remarks: Larvae use their heavy mandibles to close entrances of burrows while lying in wait for prey. Mandibles were also used to clear the burrow. Excavation pellets accumulate around the entrances of the burrows (Fig. 20A). Adults move freely on water-saturated sediments capturing prey (mainly aquatic meiofauna).

Fig. 18. Traces of the salt beetle *Bledius capra* (Staphilinidae).
A) The burrow starts with a diagonally oriented "bottle neck" and merges into a vertical lower part (burrow of a male). A and B: Resin casts.
B) Breeding chambers branching sidewards off the main shaft characterize the burrow of a female. Note also appendices filled with feces. A and B: Photograph by H.-E. REINECK.
C) Thin section showing storage of food (lumps of microbial mats) at the conversion from the "bottle neck" to the vertical shaft. Scale is 1 cm.
D) SEM-photography showing head and mouth parts of *B. capra*. Note heavy mandibles which allow the beetle to dig and clear its burrow and rip away at tough mat substrates for food. Scale is 250 µm.
E) Sheetflood (grain-flow) sediments filling a burrow of *B. capra* which was formed within a multilayered microbial mat sequence. Thin section from sediment core. Scale is 5 mm.
F) Various excavation pellets of salt beetles.

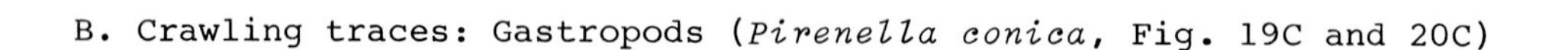

B. Crawling traces: Gastropods (*Pirenella conica*, Fig. 19C and 20C)

Description: Bilobed crawling trails on the muddy water-saturated sediment surface.

Habitats: Seawater springs with benthic macroalgal scum (*Enteromorpha* sp., probably also *Chaetomorpha* sp.). and metahaline mud flats.

Trophic relations: Grazer.
Remarks: Microbial mats do not form where the snails are abundant.

C. Feeding traces: Water beetles (Fig. 19D and 20D)

Description: Irregularly shaped holes on mat surfaces. The beetles rip away at surface substrates of submerged microbial mats. Diameter of holes made by the small hydraenid beetle *Ochthebius auratus* are 1 to 2 mm, holes made by the larger species *Enochrus* sp. are 5 to 10 mm in diameter.

Habitats: *Enochrus* sp. (salinity restricted): Marginal impondments with soft muculous and partially floating microbial mats; *Ochthebius auratus* (non-restricted by salinity): Whole mat-covered aquatic areas.

Trophic relations: Grazers.

Remarks: No indication that the occurrence of beetles is detrimental to the growing mat systems.

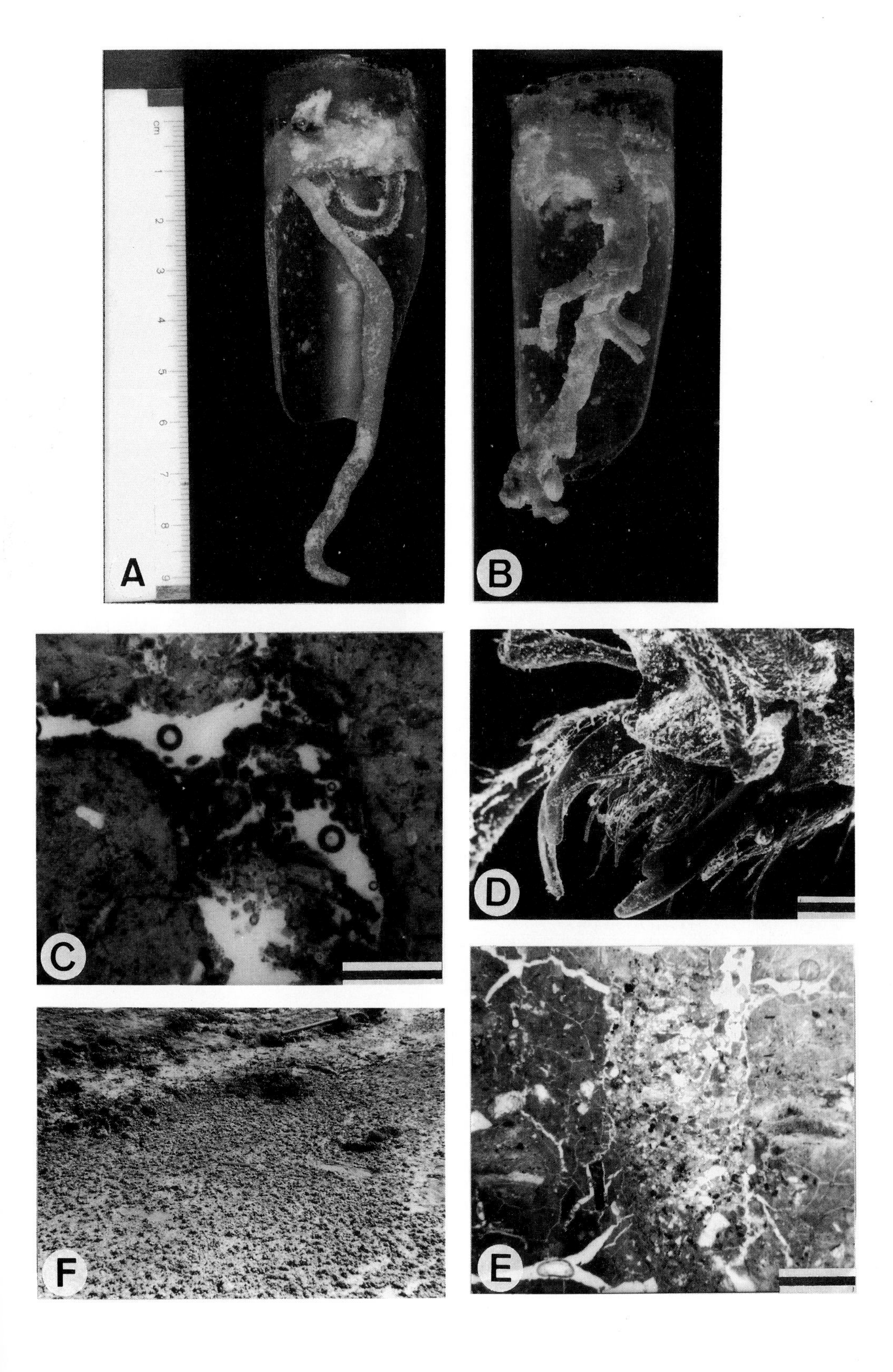
cm
1
2
3
4
5
6
7
8
9
A
B
C
D
F
E

Fig. 19. Further trace records of the Gavish Sabkha fauna.

A) Epibenthic gypsum-encrusted "chimneys" of the salt beetle *Bledius angustus*.

B) Feeding traces of *B. angustus* at the sediment-gypsum crust interface and entrance to its endobenthic retraite-burrow (arrow).

C) Bilobed crawling traces of gastropods (*P. conica*) on muddy water-saturated flats at the lagoon's margin.

D) Feeding traces in subaquous surface mats, made by water beetles (*Enochrus* sp.).

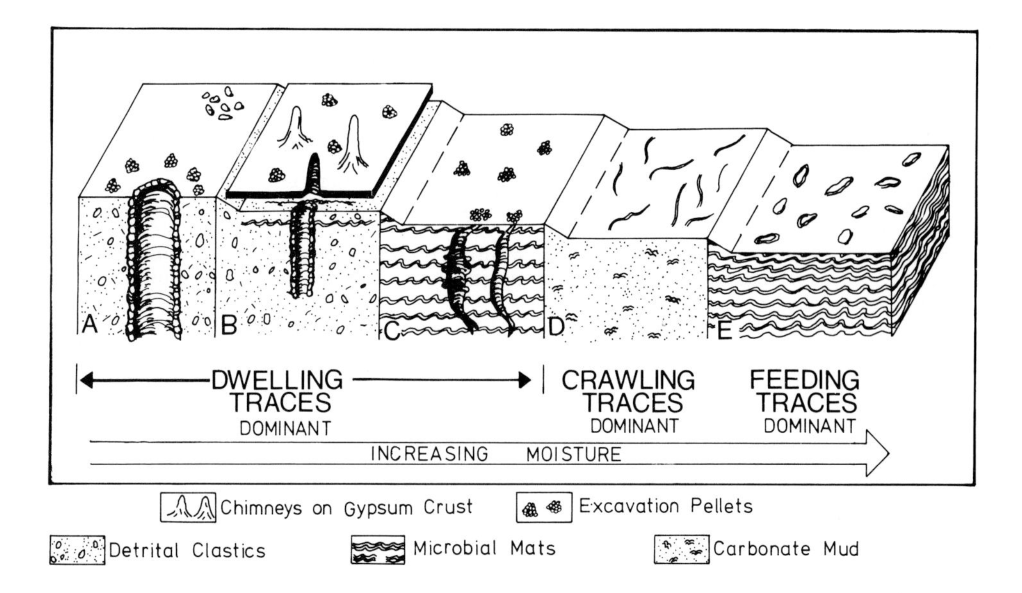

Fig. 20. The zonation of trace categories (generalized presentation). Dwelling traces are in greatest abundance at higher exposed flats, crawling traces at lower, water-saturated flats and feeding traces finally at submersed flats.
A) Burrow and excavation pellets of tiger beetle larva.
B) Burrow, excavation pellets and "chimneys" of the salt beetle *Bledius angustus*.
C) Burrows and excavation pellets of the salt beetle *Bledius capra*.
D) Crawling traces of the gastropod *Pirenella conica*.
E) Feeding traces of water beetles on a mat surface.

1. 7. 4. Environmental zonation of trace categories

Dwelling burrows predominate on the elevated, air-exposed margins around the saline basin of the Gavish Sabkha. A zonation gradient is observed between tiger-beetle larvae and the two salt beetle species *B. angustus* and *B. capra*, due to increasing moisture (Fig. 20). Crawling traces are characteristic of restricted parts of the saline mud flats and feeding traces mainly predominate on the water-covered microbial mat surfaces (Fig. 20). The gardening effect of the salt beetle *Bledius angustus* also contributes to characteristic feeding traces, which are, however, clearly distinctive from the feeding traces of the aquatic beetles. Periods of low-water levels encourage the salt beetle *Bledius capra* to migrate into deeper-lying areas and to form their characteristic bottle-necked burrows within multilaminated microbial mats (Fig. 20). These burrows clearly indicate periods of air exposure.

Fig. 21. Skeletal records of the Gavish Sabkha fauna.
A) Concentration of gastropod shells within sheetflood deposits. X-ray radiograph. Scale is 1 cm.
B) Gastropod shell with infillings of detrital clastics, embedded within a microbial mat, indicating transport by sheetfloods. Scale is 1 mm.
C) Detrital clastics within a gastropod shell showing geopetal structures. After embedding within a microbial mat, coating by microbes took place. Scale is 1 mm.
D) Outer and inner colonization of a shell fragment by microbes. Note calcified filaments of cyanobacteria on the internal shell wall. Scale is 1 mm.
E) Nodule-forming Pleurocapsalean within and outside a gastropod shell. Scale is 2 mm.
F) Soft-body remains of an ostracod, enclosed by its bi-valved carapax. Scale is 500 μm.
G) The post-mortem colonization of an empty ostracod carapax by microbes is indicated by the same filigrane network inside and outside the shell. Scale is 250 μm.
H) Dipteran pupa ton in a microbial mat implying in-situ metamorphosis. Scale is 3 mm. (B to H: Thin sections from sediment cores.)

1. 7. 5. Skeletal hard parts

Shells of the gastropod *Pirenella conica*, carapaxes of the two ostracod species and chitinous fragments of insects are common in almost all types of strata. The animals live in the Gavish Sabkha, but most are restricted to certain salinity zones (Fig. 17). The wide distribution of their skeletal hard parts even within places which lie above the salinity boundaries of the faunal populations indicate allochthonous transport which is achieved by the gravitative flow of water and sediments in the event of sheetfloods. The results of biostrationomic analyses are outlined below. The analyses focussed mainly on elaborating criteria of transport and in-situ embedding.

a) Gastropod shells: Allochthonous transport of gastropod shells is indicated by (1) concentrations of shells in debris- and mud-flow deposits showing commonly a low degree of orientation (Fig. 21A), (2) shells embedded in microbial mats and filled with detrital clastics (Fig. 21B), sometimes also showing geopetal structures (Fig. 21C). Shell fragments embedded in mats are commonly colonized after transport by microbes from the mats (Fig. 21C). Fig. 21D shows filaments in a shell fragment which are already calcified.

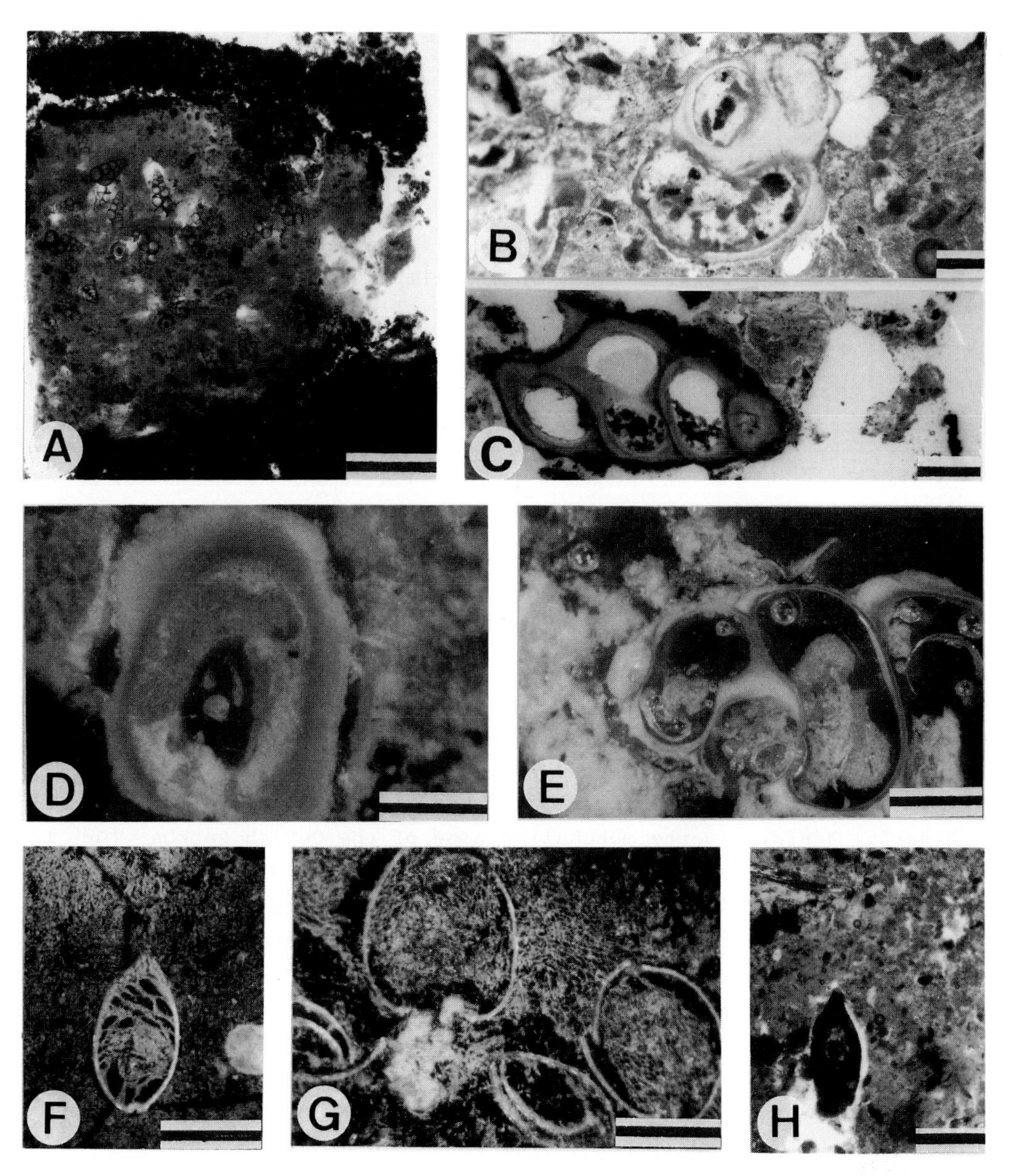

In-situ embedding: Sparse numbers of juvenile snails were found alive in the shallow water bodies where the nodule-forming Pleurocapsalean cyanobacteria mainly occur. After death and in-situ embedding the interesting nodule formation in the coating of shells by Pleurocapsalean colonies is typical (Fig. 21E). Nuclei of irregularly shaped oncoids (5 to 8 mm in diameter) consisting of gastropod shells are also reported from the lower Cretaceous of the adriatic region (TISLJAR, 1983). The suggestion of the author that the gastropods represent elements of a faunal community which lived under restricted conditions corresponds well to a biotope like the modern Gavish Sabkha.

b) Ostracods: Fig. 21F shows the soft body remains of an ostracod enclosed by its bi-valved carapax. The animal was probably associated with the microbial mat environment and after death became embedded into the decaying organic matter. Typical for the embedding substrate is the relation between bacterial decay and calcification (see chapter 1.6.4), which acts also with the soft body remains of the ostracod. By contrast, the ostracod carapax in Fig. 21G does not contain parts of the soft body. Cyanobacteria which colonized the empty shell show again that calcification immediately takes place within the embedding sediments.

c) Insects: Embedded chitinous fragments of insects which live in or close to the microbial mat environments include head capsules, fragments of extremities and pupae tons (Fig. 21H). The pupae in particular indicate that the insects chose the mat substrate for reproduction.

1. 7. 6. Grazing stress (experimental approach)

Fig. 19C gives an impression of the high abundance of the snail *Pirenella conica* in the metahaline zone of the Gavish Sabkha (up to 70 $^{o}/oo$). Possibly, mat formation is hindered here due to bioturbation and grazing, and multilaminated mats can only develop and survive under the protection of higher salinity. Laboratory experiments have shown that *Pirenella conica* can completely destratify the multilaminated mats. The experiment was carried out with a mat section which was treated with slightly hypersaline seawater (50 $^{o}/oo$). Snails were allowed to graze upon the mat section. The typical structure, resulting from the joined appearance of filamentous and unicellular cyanobacteria was destroyed as a result (Fig. 22).

Analyses of fecal pellet contents have shown, however, that many of the mat-forming microorganisms, especially unicellular cyanobacteria, are able to survive passage through the snails' intestine unharmed. These microorganisms could also participate in mat-forming processes. However, as bioturbation and grazing proceeds, microbes in the vicinity of the gastropod colonies form only diffuse aggregates which are highly enriched in unicellular types. Consequently, bioturbation and grazing result in a radical transformation, even of the community type, since certain species, often unicellular, are favored to survive, while others (mainly filamentous cyanobacteria) become rarer.

Fig. 22 A). SEM microphotograph of the undisturbed surface of a microbial mat which was sampled to study the effect of grazing. Scale is 200 µm.

Fig. 22 B). Grazing activities of the snail *Pirenella conica* led to the surface destruction of the mat. Scale is 1 mm.

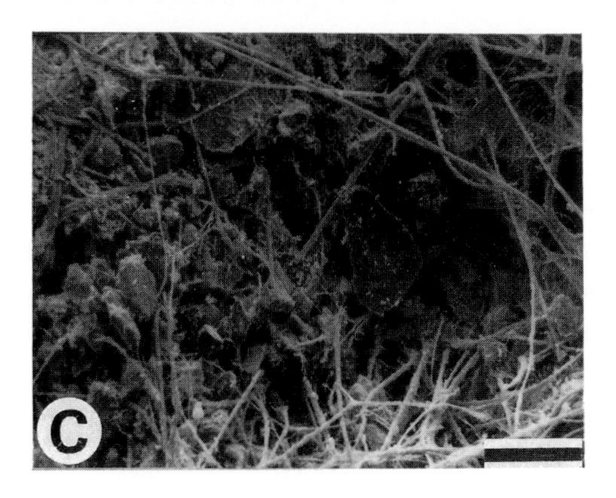

Fig. 22 C). As a result of continued grazing, the microbial mat section is destroyed and its base consisting of siliciclastic sediments is visible. Scale: 200 µm.

Fig. 23. Small-scale changes in the stratification of Gavish Sabkha sediments, drawn from thin sections from cores. All cores were sampled at the lagoon's margins. Mechanisms responsible for the change are (1) oscillations in biotopic conditions during fair weather periods such as periodic exposure to which microbes and fauna correspond (2) interruptions of the growing mat sequences by sheetfloods. For details see text.

1. 8. Modes of stratification

A well-defined change in facies types characterizes the vertical sequences of the sabkha deposits. It reflects the influence of long-lasting fair-weather conditions and short but catastrophic sheetflood events. A seasonal rhythmic (varvite-stratification) is, however, not indicated for the following reasons: (1) Fair-weather conditions with generally slower accumulation prevail for most of the year or longer. Sheet floods cause rapid sedimentation but only for a matter of hours, sometimes after a number of several years of fair-weather conditions. (2) A result of the topography is that the depression fills with fresh water on the occasion of stronger sheetfloods. Even after re-establishment of fair-weather conditions, suspended loads of fine material settle until the water is evaporated and the conditions of seepage and evaporative pumping again become established, sometimes first in early summer. (3) Climatic oscillations over several years result in sheet floods of varying energy levels.

The dynamic history is reflected in the different thickness and irregular alternation of strata types. The trend of post-event colonization of sheet flood sediments by microbial communities is repeatedly visible, for instance in core A (Fig. 23A). A further mode of fragmentation of the organic material is visible in the upper part. Such fragmentation is caused by shrinkage of the prolific microbial mats and indicates subaerial exposure. Burrowing of subaerially exposed mats by *Bledius capra* can also cause shrinkage cracks. The subaerial exposure of the upper part of the core is also indicated by the burrows of salt beetles which are unable to withstand longer periods of flooding.

Core B (Fig. 23) was taken at the saline mud flat bordering the central basin of the lagoon. The sequence shows the upward transition from regularly laminated potential stromatolites with ooids and oncoids (present-day facies type of the permanently water-covered lagoon) to a strata of the same type which is interspersed with gypsum crystals, and

Sediments from grain-flow

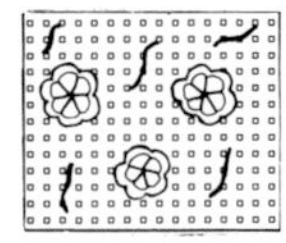
Pleurocapsalean nodules and some filaments in carbonate mud

Sediments from mud-flow

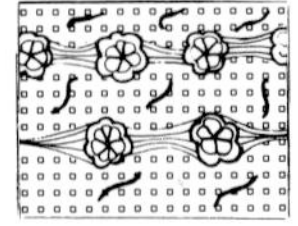
Biolaminoid arrangement of pleurocapsalean nodules

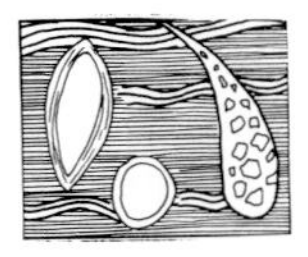
Salt beetle burrows. At right: Filled with clastics

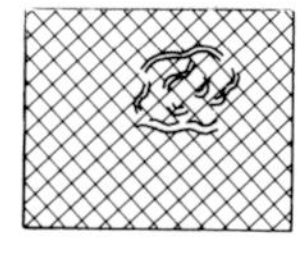
Gypsum nodule coated by microorganisms

Synaeresis-cracks and some coarser clastics in mats

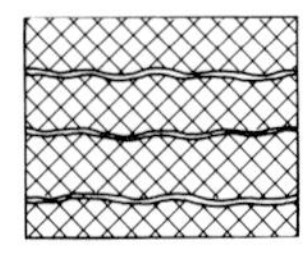
Faintly laminated sulfate

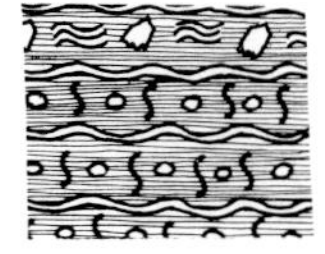
Stromatolitic carbonates with ooids and oncoids. Gypsum crystals at top

Skeletal hardparts: Top: Ostracod, Bottom: Gastropod shell

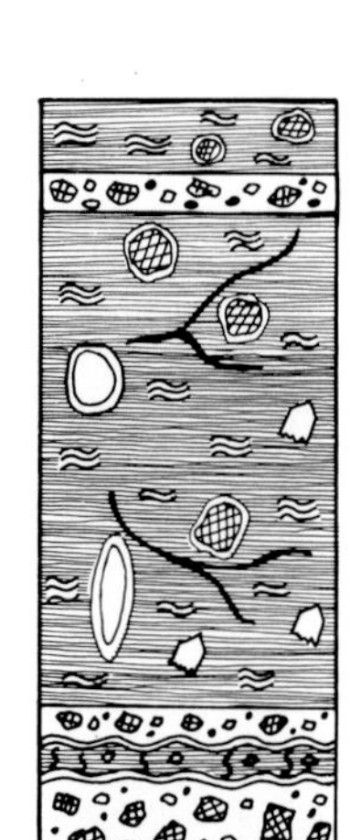

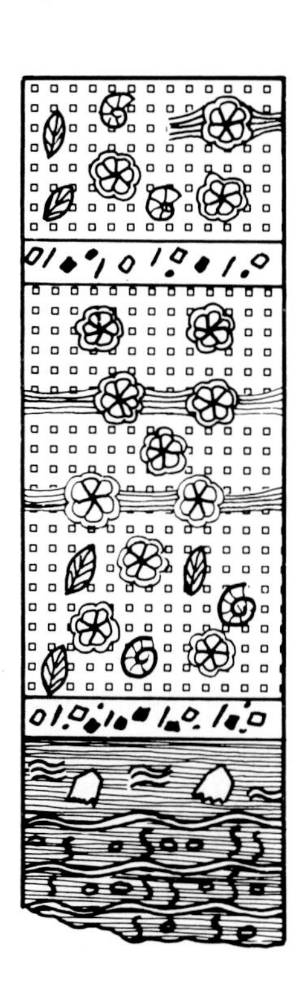

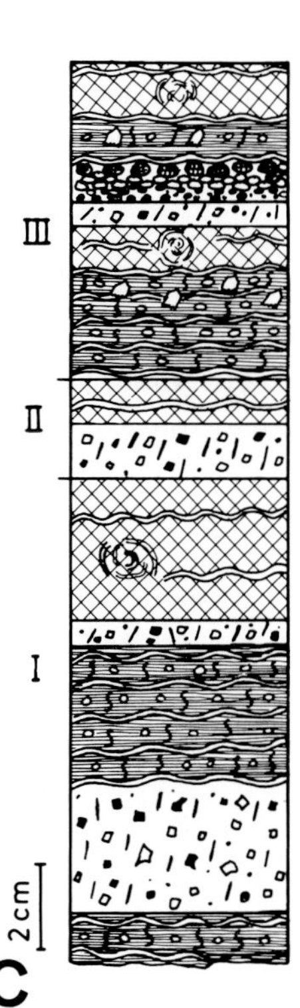

finally to a thin bed of sheetflood deposits (mud with plant detritus). The nodular biolaminoid facies of the present-day saline mud flat is established on top of the sheetflood deposition and, though also interrupted by a thin clastic flood layer, is re-established again at the surface. The core shows that ecological conditions allowing the development of the regular stromatolitic facies with ooids and oncoids were invariably absent in the growing sequence. The trend is from permanent water cover to subaerial exposure.

Core C was taken from the center-oriented saline mud flat which is characterized at the surface by gypsum precipitation. The following history of deposition can be read from the core:

Section I: A multilaminated post-event sequence of microbial mats developed on a thick sheetflood layer and was again buried by a sheetflood. Though this layer is thin it seems to have been catastrophic for the microbial community, probably due to movement of freshwater over the Gavish Sabkha. During evaporation a thick gypsum layer has formed on top and allowed re-colonization by microbes which produced faint laminated structures within the gypsum mush.

Section II: The same upward transition from sheetfloods to microbial colonization to gypsum is repeated in this section.

Section III: Finally, the evaporite sequence is followed by multilaminated microbial mats which document a longer period free of disturbance and with moderate water availability. On the surface disturbances again occur which may be the result of climatic irregularities (stronger evaporation in summer, sheetfloods in winter).

1. 9. Summary and conclusions

Grain sizes and mineralogy of sediments, community structures, standing crops, redox potentials and pH are highly correlative to the tectonically characterized terrane niveau of the Gavish Sabkha. Increasing evenness in moisture supply is realized by the inclination of the system below mean sea level. Varied physical, chemical and biological effects consequently follow (Table 6).

TABLE 6. Summary of interrelated facts which coincide with increasing moisture supply in the Gavish Sabkha

INCREASING MOISTURE SUPPLY COINCIDES WITH		
CATEGORIES STUDIED:		
A. Seawater chemistry	increasing	salinity
	increasing	Mg : Ca- ratios
	decreasing	Ca : SO_4-ratios
B. Microbial communities	increasing	diversity
	increasing	productivity
	increasing	biological self-production of laminated deposits
	increasing	intensity in sulfate reduction
C. Physicochemistry	increasing	anoxic conditions in sediments
D. Lithology	increasing	biogenic carbonate contents (mg-calcite)
E. Stromatolitic structures	increasing	regularity in the lamination of stromatolites
	increasing	abundance of ooids and oncoids
F. Faunal influence	decreasing	grazing and bioturbation intensities

To explain the increasing self-production of sediments by microbial activity (Table 6B) the convergence of (1) the specific character of the sediment-forming microbes and (2) their migrational response to undulating environmental factors has to be considered. An organism tends generelly to obtain the resource it needs. The individual cell or cell chain consists of a large portion of immobilisated compounds (sheaths, capsules, cysts and gels). Since the immobilised matter is part of the phenotype of the organism, it has to be regenerated immediately after the motile cell or cell chain has taken up a new position. This leads to a consequent increase of organic sediments as long as appropriate environmental conditions are assured.

To explain the increasing structural regularity of biolaminated deposits (Table 6E), the evenness of seasonal shifts within the permanently water-covered environments may be emphasized which allow the organisms to correspond with changing gradients of environmental stimuli, such as light intensities, salinity, moisture, pH and reduction-oxidation potentials.

Factors changing the light conditions in the Gavish Sabkha are - apart from day-night cycles - light channeling by water-cover, salinity and particles suspended in the water. Salt crusts at the sediment/air interface and oversedimentation (by eolian or fluviatile transport) also change the light intensity. All these factors can fluctuate seasonally or aperiodically. Qualitative change can stimulate subsurficial populations to override other topmats. Thus the variety of slightly shifting environmental factors significantly determines the "depositional dynamics" of stromatolitic growth-bedding (sensu PETTIJOHN & POTTER) in the Gavish Sabkha.

The lateral sequence of facies type 1 to 4 is actually the expression of arid fair-weather conditions which last at least 9 to 10 months a year. Interlayered bedding of terrigenous detrital clastics, evaporites and biogenic sediments provide clues about

1. the climatic setting of the depositional environment of the Gavish Sabkha: Long-lasting stable arid periods without rainfall are suddenly interrupted by storm events,

2. its geomorphic state: depression, no surface connection to the sea, gradually filled by mainland-generated transport protruding seawards.

"Das subfossile Riff am Strand lehrt uns, wie jene negative Bewegung des Strandes bis in die jüngste Vergangenheit hinein fortdauert."
(JOHANNES WALTHER, 1888)

2. THE SOLAR LAKE - IMPORTANCE OF SMALL TECTONIC EVENTS

(GULF OF AQABA, SINAI PENINSULA)

2. 1. Introduction

The Solar Lake lies about 200 km north of the Gavish Sabkha (Fig. 2a). According to a stratified saline water body, temperature in lower strata reach up to 65 ^{o}C. This "accumulator of solar energy" (KALECSINSZKY, 1901) is characterized by a well-defined annual cycle: Stratification prevails between September and June, monomixis between July and August. Overturn takes place when seepage inflow from the adjacent Gulf can no longer compensate for the water loss by insolation.

The Solar Lake forms a basin reaching a maximum depth of 5 m below mean sea level. Microbial mat sequences cover the shelf to unusual thickness. From these sequences we can read that the sun-energy promoted cycle of the Solar Lake has been operating for nearly 2,000 years (KRUMBEIN & COHEN, 1974, COHEN et al., 1977a, FRIEDMAN, 1978). Before that time the area was a shallow back-barrier lagoon similar to the Gavish Sabkha. The change was caused by a sudden subsidence of the central part of the lagoon resulting from fault movement. This event produced a deeper basin and consequently a water body of lower surface-to-depth ratio compared to the Gavish Sabkha.

The purpose of this chapter is to undertake a case study of changes in facies resulting in regularly laminated stromatolitic carbonates overlying a sabkha-type deposit. Such facies change can be identified in several stratigraphic records (see for example in the Permian Zechstein (PZ3) sequence mentioned in Part III of this volume). The chapter is based (1) on our own studies of core material from the Solar Lake, sampled parallel to the Gavish Sabkha material (for methods see chapter 1 of this Part), and (2) on studies of existing comprehensive literature about this lake.

2. 2. Locality and previous work

Coastal plains typical of the southern part of the Sinai coast of the Gulf of Aqaba are absent in the northern region. Mountain sides of the Sinai rocky desert flanking the Gulf slope steeply towards narrow fringing reefs and frame some semicircular sandy beach depressions. The Solar Lake, which is about 140 m long and about 65 m wide, is located in one of these semicircular depressions (Fig. 24). It is fed by seawater seeping through a completely closed gravel bar which is 60 m wide and 3 m above mean sea level. The semicircular terrane, protected from wind by the mountain ridges and from flood recharge and surf by the bar, recalls an amphitheatre on whose "stage" (the sheltered center), sun, water and organisms perform their sediment-accumulating ritual. Moreover, the effects of flashfloods which play an important role in the unprotected Gavish Sabkha are made milder, since the Solar Lake is separated by its southern mountain flank from a larger sheetflood basin where the most part of the sediment-laden freshwater is stored. Only a narrow overflow passage between the coastal bar and the southern mountain flank leads into the Solar Lake.

Fig. 24. The Solar Lake at the shore of the Gulf of Aqaba, Sinai Peninsula. The lake forms a 5-m-deep basin within a semicircular depression which is flanked by steep mountains sides. Partial exposure of the shallow shelf of the lake indicating summer situation. Lower part: Coastline with fringing reef. (After GERDES et al., 1985.)

This lake may be among the best investigated in the world. Its limnology was studied by POR (1968, 1969), NEUMANN (1968), ECKSTEIN (1970) and COHEN et al. (1977a). FRIEDMAN et al. (1973) surveyed biogenic carbonates within the biolaminated deposits (see also FRIEDMAN, 1978). KRUMBEIN & COHEN (1974) and KRUMBEIN et al. (1977) studied the biogenic and abiogenic sediment accumulation and depositional history.

COHEN et al. (1977b, c), KRUMBEIN & COHEN (1977) and COHEN (1984) described the microbial communities in terms of physiological data, primary production, mat formation and lithification. EHRLICH (1978) dealt with the diatom flora and POR (1975), DIMENTMAN & SPIRA (1982) and GERDES et al. (1985) studied the fauna. Biogeochemical data were provided by BOON (1984), AIZENSHTAT et al. (1984) and COHEN et al. (1986). GIANI et al. (1984) studied the methanogenesis within the biolaminated deposits. AHARON et al. (1977) calculated isotope ratios and gypsum precipitation from evaporation experiments.

2. 3. Bathymetric zones and limnologic cycle

Three bathymetric zones can be distinguished: Shelf, slope and bottom. The shelf is a gently inclined, 20 to 25 m wide area which is surrounded by a sandy shore bare of vegetation and with beach-rock outcropping. The slope starts at about 1.50 m water depth. It consists of an initially gentle part which merges in about 2.50 m water depth into a steeper part running down toward the bottom of the basin.

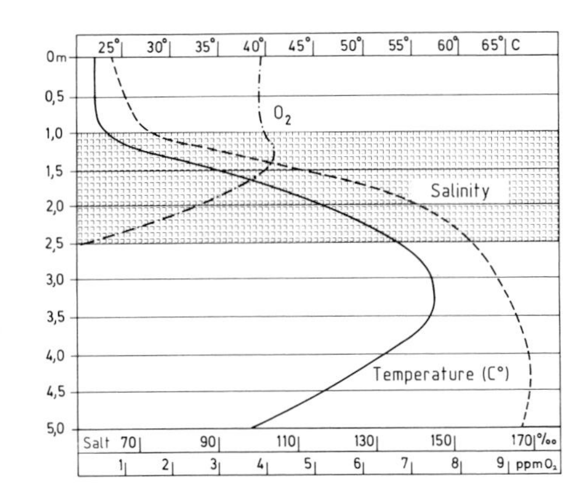

Fig. 25. Vertical distribution of temperature, O_2 and salinity within the lake during winter stratification (modified after Gerdes et al., 1982).

From September until May the lake's water body is stratified (Fig. 25). Even snorkling at that time is hard since the metalimnion begins a few dm below the water surface. In between a depth ranging from -1 m to -2.50 m below water table, temperature increases from 25 to 50 °C, salinity from 70 to 150 ‰ and oxygen diminishes to less than 1 ppm. In the upper hypolimnion, temperature can reach 65 °C and salinity 180 ‰. The hypolimnion is completely anaerobic. Relative to the meta- and hypolimnion, the epilimnion is cool at 20 to 25 °C. It is rich in oxygen and the salt concentration lies between 70 and 80 ‰.

Since the water body is four to five m deep and seepage continues throughout the year, even evaporation exceeding the inflow rate of seawater does not lead to drying out in summer, but with increasing evaporation and water loss, the mesothermy of the water body becomes unstable. In falling below the stability point, a sudden mixing up of the previously stratified water bodies takes place. At this point, the salinity of the entire water body lies between 160 - 180 ‰, and temperatures of 27 °C prevail throughout the water column (COHEN et al., 1977a). The period of holomixis lasts from 4 to 13 weeks, mainly between July and September.

2. 4. Sub-environments and facies types

2. 4. 1. The shelf

a) Biotopic conditions

On the relatively flat shelf the effects of seasonal fluctuations in the lake's water level are considerable. COHEN et al. (1977a) measured a change of 1.4 m over the year. In late summer, about 50 % of the shelf is exposed (Fig. 24), and water depth in the lower part of the shelf is 10 to 15 cm. During this low-water period springs of seawater are above the original water level. With the replenishment of water in early fall, seawater springs and finally the whole shelf becomes water-covered and water depth on the lower shelf again reaches 1.50 m.

Light intensity and salinity values change with the seasonally oscillating water table. Within the stratification period, salinity ranges between 50 and 80 ‰, while it is about 180 ‰ within the summer mixing period. Since the shelf is not supplied with the hot and anaerobic brine of the deeper basin, water temperatures do not reach higher values than 30 °C.

b) Main mat-building organisms

The shelf mats are made up predominantly of cyanobacteria. Filamentous species (*Oscillatoria limnetica/Phormidium hendersoni*, *O. salina*,

Lyngbya sp., *Microcoleus chthonoplastes*) and unicellular forms (mainly of the genera *Synechocystis* and *Synechococcus*) are abundant. Unicellular cyanobacteria of the Pleurocapsalean type (multiple fission) are far less abundant, hence there is no formation of nodular cell aggregates characteristic of the Gavish Sabkha. Diatoms are well recorded (EHRLICH, 1978), the genera *Nitzschia*, *Amphora* and *Navicula* predominating both in species and specimen abundance in living surface mats. The diatoms, however, generally are not preserved in older sections of the microbial mats.

With increasing light together with less or no water covering in summer, unicellular cyanobacteria (*Synechococcus*, *Synechocystis*) and diatoms predominate in the top layers. These organisms form protective pigments (carotinoides) which give the surface a brown-yellow colour.

In fall, winter and spring, at higher water levels and thus slightly more reduced light conditions, filamentous cyanobacteria (*Microcoleus chthonoplastes*, *Oscillatoria limnetica*, *O. salina*) override the coccoid/diatom community. The surface appears blue-green at this time.

c) Mat construction and microfacies

As in the Gavish Sabkha, the biolaminated sediments of the shelf of the Solar Lake consist of sets of dark L_h- and light L_v-laminae, the L_v usually 4 to 8 times thicker than L_h (Fig. 26A). The light L_v-laminae are characterized by a high content of extracellular slime and a variety of irregularities (decaying cell clusters, intraclasts, voids and bubbles). The dark L_h-laminae form usually condensed horizons built up of horizontally oriented ensheathed bundles of filamentous cyanobacteria. The partition of L_h-laminae is common, giving rise to the formation of eye-shaped lenses (Fig. 26B, D). The lenses enclose material from the light L_v-laminae. Thin section studies and scanning electron microscopy show that miniscule voids, probably derived from gas bubbles, are initially colonized by microorganisms (Fig. 26E). Carbonate precipitation takes place mainly in the light layers and the eye-shaped lenses where ooids and oncoids occur in great abundance. They lie very close together within the eye-shaped lenses (Fig. 26C, D), others within the widely spaced L_v-laminae are more dispersed (Fig. 26F).

In particular where oolites and oncolites in the fossil record are devoid of a biostromate lamination which could imply in-situ formation

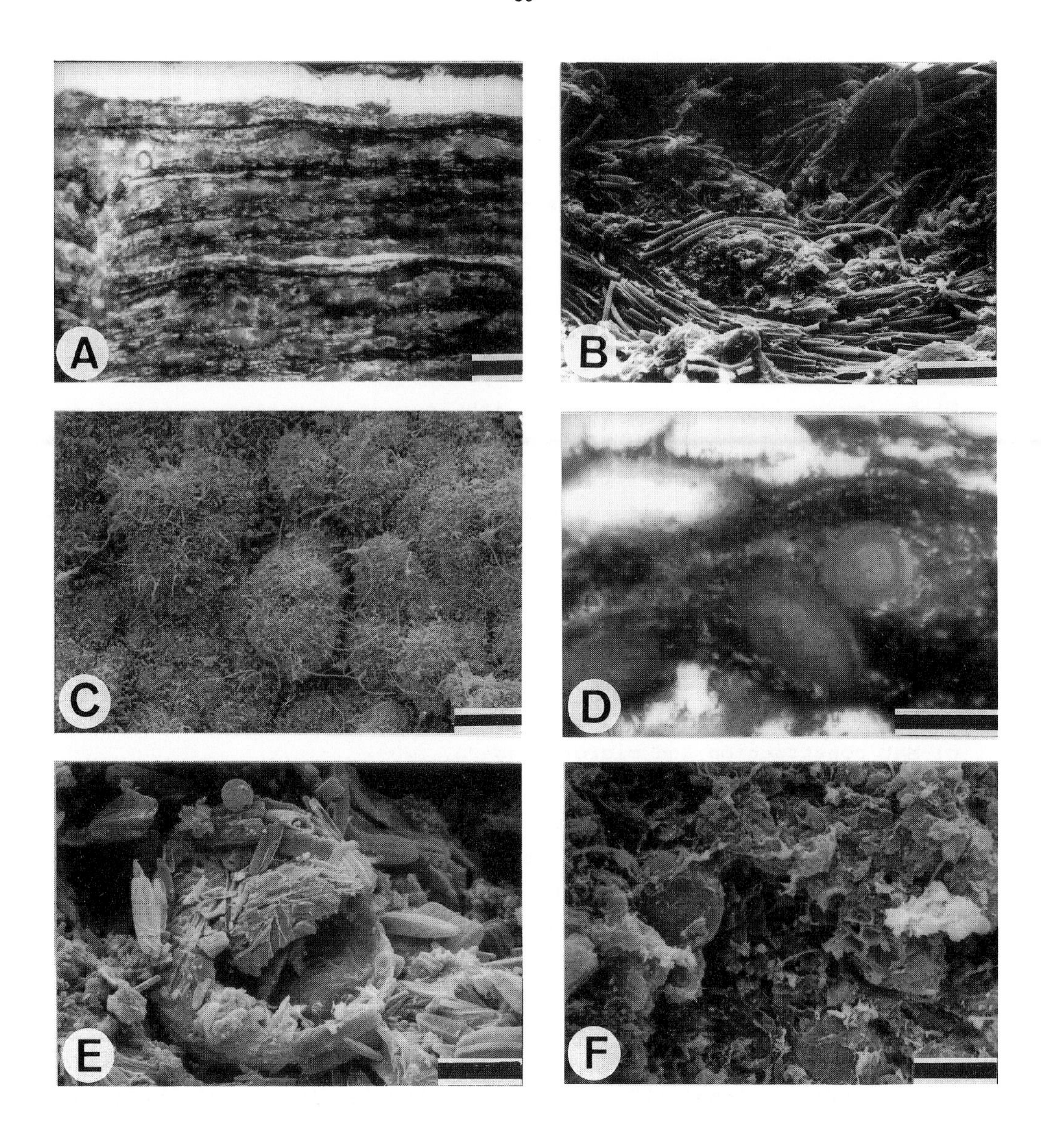

of the grains, their close arrangement within eye-shaped lenses may indicate the formational site which had formerly a microtopography similar to the Solar Lake microbial mats (see for example the Minette ooids and oncoids; Fig. 43A, B and C).

Fig. 26. Microfacies characteristics of the shelf deposits.
A) Lamination including dark L_h- and light L_v-laminae. Scale is 1 mm.
B) Ensheathed filament bundles of the L_h-lamina forming an eye-shaped lense. The lense contains decaying organic matter and polysaccharide slimes and initial nucleation of coated grains (see also C and D). Scale is 25 µm.
C) "Coated grains": Bedding plane concordant view showing the coating of ooids and oncoids by filamentous microorganisms. Scale is 50 µm.
D) Vertical section showing layer with ooids and oncoids. Note arrangement of the grains within the eye-shaped lense. Similar arrangements of coated grains in rocks (compare Fig. 43C) may imply their formation in a similar microbial mat environment. Scale is 250 µm.
E) A hollow sphere (bubble?) within a L_v-lamina, coated by diatoms and coccoid unicells. The wall of the sphere is stabilized by polysaccharide slime (internal view). Scale: 25 µm.
F) Dispersed arrangement of miniscule spheres within the decaying organic matter and mucus of a L_v-lamina. Scale: 10 µm.

d) Stratification and vertical extension of the shelf mats

Particularly on the lower shelf, the biolaminated deposits form over a vertical distance ranging from 1.0 to 1.20 m. The whole sequence contains the stromatolitic carbonate facies type with L_h- and L_v-laminae, ooids and oncoids already described for the Gavish Sabkha. The seasonal cycle controls a varvite-like formation of alternating dark and light laminae (Fig. 27). From time to time, the regularly laminated vertical succession is interrupted by more crenulated sections, often followed or underlaid by thin strata of mud- or grain-sized clastic sediments. L_v-laminae within the crenulated sections are commonly thicker than within the regularly laminated sequences. Vertical sections show increasing numbers of small bags, pockets and eye-shaped lenses. Synaeresis cracks, extraclasts and gypsum crystals are common which also penetrate and interrupt the thin and fragile L_h-laminae. In summary, the varvite-like formation of the shelf deposits changes from time to time to unconformities and turns back again to the regular microstratification.

These patterns indicate climate deviations from the norm, probably occurring secularly and taking place after a sequence of regular seasonal change. The climate deviations are accompanied by hotter and drier summers, consequently by higher evaporation rates, increasing salinity and light conditions, and higher rainfall in winter, which causes sheetfloods.

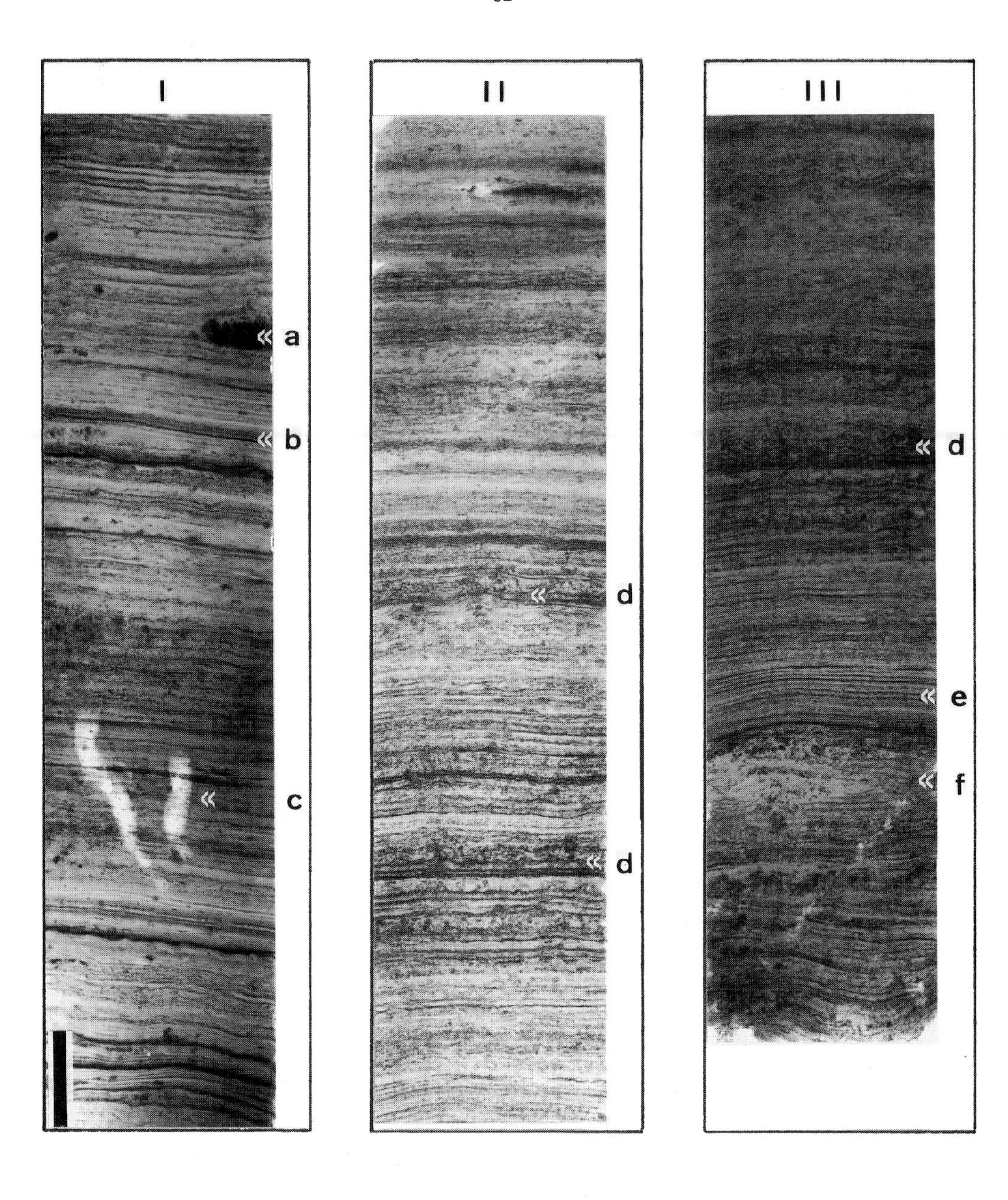

Fig. 27. Deposits of the Solar Lake shelf. X-ray radiographs of sections of an originally 80 cm long core. From top (section I) to bottom (section IV) the sequence shows the stromatolitic carbonate facies with typical light-dark interlaminations, ooids and oncoids. Individual constituents providing clues on the environment are listed on the next page. X-ray radiographs by H.-E. REINECK. Scale is 2 cm for all core sections.

Legend to **Fig. 27** continued:
Core section I:
a) Intraclasts of varying size and shape, resulting from the liberation of mat fragments from the lake's deeper slope, drifting in the water and depositing at the water rim on top of the shallow shelf mats.
b) Sheetflood-derived mud layers 5 to 10 mm thick.
c) Burrows of the salt beetle Bledius capra in 13 to 16 cm depth indicate a former period of subaerial exposure.
Section II:
d) Deviations from the normally stratiform to a more crenulated pattern of the dark L_h-laminae, accompanied by the aggregation of ooids and oncoids within the "valleys". The crenulation comes from micropinnacles which form at mat surfaces as a reaction of the mat-community to increasing light and salinity (see also Fig. 41F).
Section III:
e) Regularly alternating dark and light laminae where ooids and oncoids show a chain-like arrangement due to the low extension of the light-colored L_v-laminae. These sequences indicate growth under a slightly higher water table and a lower impact of the seasonal change in light intensity.
f) Widely spaced light-colored L_v-laminae with clustered and dispersed arrangements of ooids and oncoids. These as well as the micropinnacles indicate climate deviations from the average, accompanied by increasing salinity and light conditions.
Section IV:
g) (sediment depth averages 75 cm at this point). Compound ooids and oncoids merging into chain-like arrangements, due to increasing compaction and dehydration of the light-coloured hydroplastic L_v-laminae.

Light and salinity oscillations in seasonal and also secular rhythms obviously are important control mechanisms. Much more polysaccharids accumulate under higher salinity and irradiation, a fact already observed in the Gavish Sabkha and here repeated. Another aspect that can be directly entangled with the climatic unconformities is the morphology of the mat surface itself. The mats of the lower Solar Lake shelf restratify themselves to micropinnacle formation under increasing salinity and light. The pinnacles are conical buildups which reach about two to three millimeters upwards from the stratiform mat base. Even the *Microcoleus*-dominated L_h-lamina in this case follow the formation of the pinnacles. In vertical sections, the pinnacle formation favors the crenulated structure, the formation of bags, pockets and eyes between the pinnacles. Ooids and oncoids here represent veritable clusters (Fig. 26, see also Fig. 27, marked section d of core section II).

2. 4. 2. The slope and bottom

a) Biotopic conditions

Effects of water level change are no longer important where the water deepens at the slope. Parts of the slope are, however, beyond the epilimnion and thus are greatly affected by the steady increase of temperature and salinity, the decrease of oxygen and the increase of H_2S. Only after the turnover in late summer are these environmental factors moderated for a short period.

Fig. 28. Thin section of the crust at the Solar Lake's bottom showing layered accretion of elongated gypsum crystals interlaminated with faint microbial mats (dark laminae). Scale is 2 cm.

b) Mat-building organisms and mat construction

The initial slope community is similar to that of the shelf. The mats tend to form soft flocculous fabrics which partially drift in the water and are subject to rapid decay. With increasing depth towards the bottom, sulfur-dependent, other anoxy-photobacteria and anaerobic bacteria increase in number. *Oscillatoria limnetica* is the only species of cyanobacteria which remains common, even at the bottom where temperatures are high and completely anoxic conditions prevail. Flocculous mats undergoing rapid anaerobic decay cover the lower slope and the bottom of the lake (KRUMBEIN et al., 1977). The internal fabric of these sediments shows large gypsum crystals and faint interlaminated strata of microbial mats (Fig. 28).

2. 5. Lithologic and ichnologic framework

2. 5. 1. Clastic compounds

That the present-day Solar Lake is nearly devoid of strong disturbances by the wind, the sea and rain-derived flashfloods can be read from its depositional record (Fig. 27). Grainflow- and mudflow-derived sediments enter the lake at less important rates than in the Gavish Sabkha. Mud layers intercalated in biogenic deposits usually range between 0.5 and 2 mm thick. Cores from a S - N transect crossing the shelf show that mud layers diminish in thickness with increasing distance from the overflow passage which is between the coastal bar and the southern mountain flank (KRUMBEIN & COHEN, 1974).

2. 5. 2. Evaporites

The almost entirely biogenic shelf sediments are free of gypsum layers, due to bacterial sulfate reduction. Subaqueous gypsum precipitation is reported by KRUMBEIN & COHEN (1974) and KRUMBEIN et al. (1977): (1) A gypsum crust forms below a surface mat on the slope at a depth of about 2.5 m. At this level, seawater under artesian pressure enters the lake and supplies additional oxygen. (2) Deeper parts of the slope are covered by a hard crust of gypsum and carbonate, precipitating from the supersaturated brines (Fig. 28). Temperatures as high as 40 to 50 °C and minimal sulfate reduction at low organic matter concentrations favor high gypsum accumulation rates.

2. 5. 3. Ichnologic patterns

A comparative study of the fauna of the Solar Lake and the Gavish Sabkha was carried out by GERDES et al. (1985d). The study has shown that species composition, topographic moisture- and salinity-related zonation in both environments are similar, particularly the predominance of coleoptera (POR, 1975). Ichnological patterns in the Solar Lake are mainly due to the staphilinid beetle *Bledius capra* (member of the wetland fauna). When the lake's shelf is water-covered (winter), the beetles live on the shoreline and burrow in sand-filled cracks and pockets of the beachrock. When the waterline of the lake starts to retreat in spring or early summer, the beetles migrate along the humidity gradient into the gradually air-exposed shelf area and form abundant bottle-neck shaped burrows within the mats (Fig. 29). Traces appearing in buried strata (Fig. 27) indicate that the deposits have been subject to exposure.

Fig. 29. Horizontal section of air-exposed shelf mats of the Solar Lake showing abundant burrows of the salt beetle *Bledius capra*. Summer situation, where retreating water line attracts the beetles to migrate from the sandy shore into the microbial mat-covered shelf area. Scale is 2 cm.

Hydrophilid beetles belonging to both stenohaline and hypereuryhaline aquatic fauna occur in numbers in the Solar Lake and graze on the mat surfaces, leaving feeding traces as already indicated for the Gavish Sabkha (see Fig. 19D). Meio- and microfaunal elements (ostracods, copepods, nematodes, turbellarians and protozoans) are abundant, as in the Gavish Sabkha; however, the gastropod *Pirenella conica* which is important in the Gavish Sabkha in terms of grazing and hard part supply is completely absent in the Solar Lake.

2. 6. Summary and conclusions

2. 6. 1. Occurrence of facies types compared to the Gavish Sabkha

Grain-flow supported siliciclastic biolaminites (facies type 1) and nodular to biolaminoid carbonates (facies type 2), both occurring in the Gavish Sabkha, are absent in the present-day Solar Lake deposits whereas stromatolitic carbonates with ooids and oncoids (facies type 3) are well developed. Biolaminated gypsum, which characterizes the amphibious rims of the Gavish Sabkha lagoon (facies type 4), forms in the Solar Lake subaquatically at the bottom of the basin and comprises another community type comparable to the amphibious areas of the Gavish Sabkha.

An informal overview of the differences mentioned here is given in Table 7. Reasons for these differences are discussed below:

(1) The Solar Lake apparently does not provide geomorphological conditions suitable to develop facies type 1. The coastal bar is upheaved by on-shore directed wave energy and does not merge with the hinterland as is the case with the circular depression of the Gavish Sabkha. Thus sheetflood-derived grain flows are not able to run along the plateau of the bar and down-slope to form sandlobes and to bury mats. Thus the horizontally flat l_h-mats sandwiched between grain-sized siliclastic sediments do not occur within the Solar Lake area.

(2) Sub-environments which favor the nodular to laminoid carbonate facies are also apparently not evident in the Solar Lake. In the Gavish Sabkha, these are the gently inclined, water-saturated and schizohaline mudflats adjacent to the lagoon. The same facies type is

TABLE 7. Existence of facies types of the Gavish Sabkha in the Solar Lake

	GAVISH SABKHA	SOLAR LAKE
FACIES TYPE 1: L_h-type biolaminites sandwiched between siliciclastic sand-sized sediments	indicating grain-flow deposits sheetflood derived (coastal bar slope)	non-existent
FACIES TYPE 2: Nodular to biolaminoid carbonates	indicating schizohaline conditions of water-saturated sediments (bar-oriented mudflats)	non-existent
FACIES TYPE 3: Stromatolitic carbonates with ooids and oncoids	indicating hypersaline shallow water conditions (lagoon)	indicating hypersaline shallow water conditions (shelf)
FACIES TYPE 4: Biolaminated gypsum	indicating hypersaline subaerial/ amphibious conditions (rims of lagoon)	indicating hypersaline/ hyperthermal subaquatic conditions (bottom of the lake)

found in mudflats surrounding the Ras Muhamed pool which lies on the southernmost tip of the Sinai Peninsula (SNEH & FRIEDMAN, 1985). This environment is very similar to the Gavish Sabkha.

(3) Though vertically more extensive, the shelf mats of the Solar Lake complete the picture already painted of the Gavish Sabkha shallow-water mats. This facies type is also repeated in the permanently water-covered shallow depression of the Ras Muhamed pool (SNEH & FRIEDMAN, 1985). This environment, as well as the Solar Lake, lacks the central elevation produced by active evaporative pumping characteristic of the Gavish Sabkha. The development of regularly interlaminated carbonates, ooids and oncoids in areas under the following environmental conditions is common to all three environments: hypersalinity, shallowness of

water, lowest level energy, and evenly oscillating patterns of water depth, salinity and light throughout the year.

The development of stromatolitic layering in all these environments without any oversedimentation has led us to avoid the definition of stromatolites as being products of sediment-fixing, sediment-binding and/or sediment-precipitating activities of microorganisms (AWRAMIK et al., 1976; BUICK et al., 1981), which would imply that physical and chemical sedimentation processes are neccessarily needed for stromatolitic buildups.

2. 6. 2. Time intervals recorded in stromatolites

Deducing annual, seasonal or even diurnal periodicity from the characteristic millimetre-scale lamination of stromatolites neccessitates some precautions even more if laminae which at first glance look massive are internally finely laminated (PARK, 1976). The biogenic sediments of the Solar Lake may serve as an example.

Core material at 1.20 m sediment depth yielded a ^{14}C age of 2400 years B.P. (KRUMBEIN et al., 1977). The overall growth calculated from this data is 0.5 mm per year including a pair of light and dark laminae. These develop without the aid of sedimentation as the result of two different kinds of microorganisms overriding themselves in order to find the most appropritate site under seasonal changing conditions. They may thus represent seasonal rhythmites. However, even the pure biogenic lamination of the Solar Lake shows some sort of irregularities. The main factor responsible is the climate oscillation. Series of years with very regularly undulating environmental conditions are abruptly or gradually followed by those with weather unconformities, including breaks in biogenic sediment accretion on the one hand, higher precipitation rates of evaporites and sheetflood sedimentation on the other hand. There remains thus a degree of uncertainty in using the obviously rhythmic lamination as a time scale. Furthermore, a reduction of the wet organic matter by processes of bacterial degradation, transformation to carbonate minerals, compaction and dehydration in the order of about 70 % may be realistic to assume (PARK, 1976; KRUMBEIN et al., 1977). The annual accretion rate would then comprise only 0.15 mm for one couplet of seasonal rhythmites. This would result in a micro-interlayered bedding representing 6 to 7 years within one single millimetre. According to these values, regularly biolaminated sequences developing during a period of very regularly undulating environmental conditions may at first glance look like one single layer and its interlayering with unconformity layers would imply periodicity on the millimetre-scale.

2. 6. 3. Importance of small tectonic events

The chief concern of this chapter has been to demonstrate the record of change from a shallow-water environment into a subsequently deeper basin. In chronological order there have been four stages in the development of the Solar Lake (1) a stage open to the sea, comparable to fjord-like embayments which occur adjacent to the Solar Lake bight, (2) a semi-closed stage due to the gradually forming coastal bar, (3) the completely closed stage and finally (4) the subsidence-derived deep basin. Coring and radiocarbon dating of organic sediments provided evidence that the change from the open to the semi-closed stage happened in the time between 4644 +/- 555 and 3378 +/- 172 B.P. (COHEN et al., 1977a, KRUMBEIN et al., 1977; FRIEDMAN, 1978). The process of bar-closing may have brought about conditions as in the Gavish Sabkha. Below the 1.20-m-thick biolaminated deposits on the Solar Lake shelf, carbonate mud occurs in which shells of the gastropod *Pirenella conica* are abundant (FRIEDMAN, 1978). This species is absent in the present-day Solar Lake area whereas it is common in various shallow-water environments along the Gulf coast, including the metahaline parts of the Gavish Sabkha and also of the Ras Muhammad Pool (SNEH & FRIEDMAN, 1985).

During the semi-closed stage, the center of the lagoon which is now just below the bottom of the Solar Lake was floored by microbial mats. It was surrounded by mud flats where slightly hypersaline conditions or even normal seawater salinity allowed colonization by *P. conica;* The gastropod zone was separated from the microbial mat zone, indicating that a salinity barrier was established which hindered the grazing of gastropods on the microbial mats. This as well as the shallowness of the former system is similar to the present situation of the Gavish Sabkha and to the Ras Muhammad Pool.

At about 2490 +/- 155 B.P. the lagoon closed off completely. *P. conica* gradually disappeared, and shallow-water cyanobacterial mats established throughout the lagoon. These conditions, lasting about 555

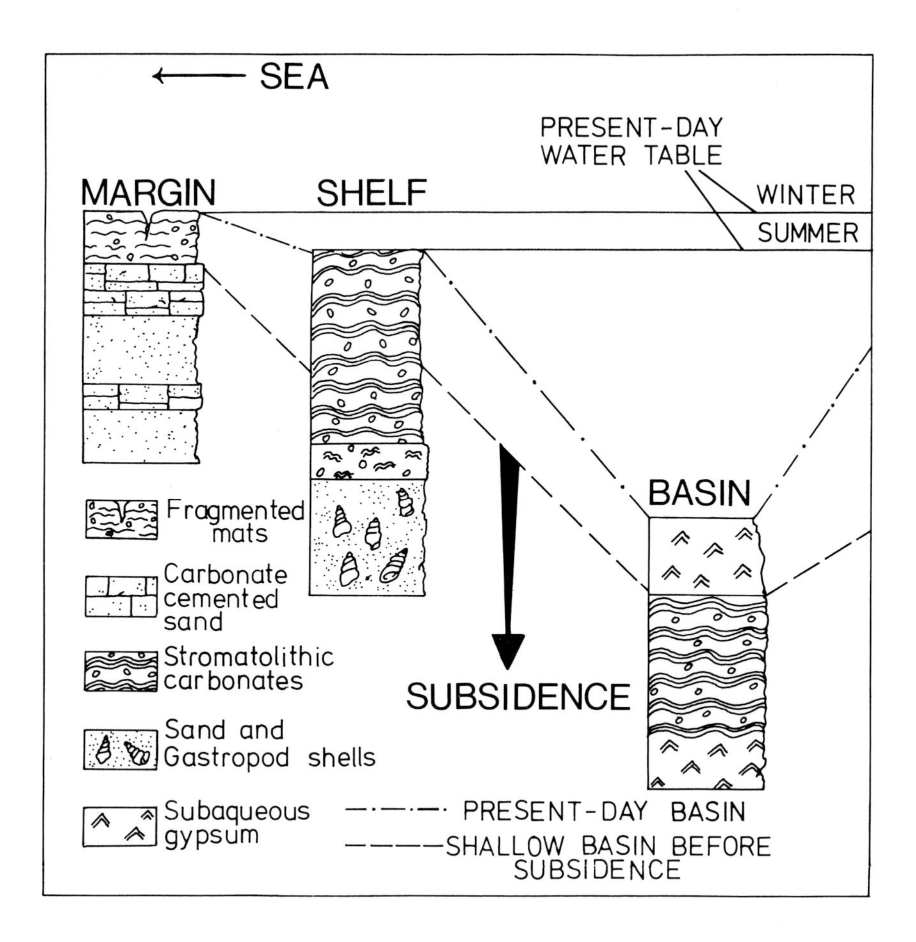

Fig. 30. Schematic presentation of facies changes caused by a fault movement which formed the deep Solar Lake basin and its limnologic cycle. Shelf: Regularly laminated microbial mats with ooids and oncoids overlying carbonate mud and sand with gastropod shells (*P. conica*). Bottom: Subaqueous, faintly biolaminated gypsum overlying regularly laminated stromatolitic carbonates which formed before the subsidence. Drawings of core successions not to scale, modified after KRUMBEIN et al. (1977).

years, came to the end due to a fault movement which caused the subsidence of the central part of the lagoon. With a deep basin established the limnological character changed. These conditions prevail up to the present and favor the shelf mats growing to extraordinary thickness while at the bottom of the lake only a few specialized microorganisms are allowed to thrive and to interfere with gypsum precipitation. Thus the change from a peritidal sabkha-type environment into a subsequently deeper mesothermic, monomictic lake is recorded within (1) the shelf sequence where carbonate mud containing "cerithid" gastropods merges into regularly laminated stromatolitic carbonates and (2) the bottom sequence where regularly laminated stromatolitic carbonates merge into faintly biolaminated gypsum (Fig. 30).

That the island is no firm land, but the peak of a mile-long sand-drift; that thus the sand here belongs spectacularly to the forces of nature - all this became more and more to be a basis of our experience (M. LUSERKE, 1957).

3. VERSICOLORED SILICICLASTIC TIDAL FLATS

(MELLUM ISLAND, SOUTHERN NORTH SEA)

3. 1. Introduction

The examples of the Gavish Sabkha and the Solar Lake show that with the aid of gently oscillating environmental stimuli (water potential, salinity, temperature) within an otherwise quiet water environment laminated sediments of purely biogenic origin develop. Early calcification within the slowly growing biogenic sequences increase the internal stability. Physical sedimentation on the other hand is more accidental than an aid to growth of biolaminites. By contrast, the rise of biolaminated deposits described in this chapter owes much more to physical sedimentation. To show pictorially the importance of wind-borne sedimentation, R. RICHTER (1926) used the term "German Sahara" when he described the common sand-drift at Mellum Island.

The "versicolored" high tide flats of Mellum Island represent (1) microbial mat environments of higher latitude (53° 43' N) than the Gavish Sabkha and the Solar Lake; (2) their lithologic character is purely siliclastic; (3) they are in open contact with the sea. Aim of the paper presented here is to describe the facies and to interpret it by ecological and sedimentological data obtained from field and laboratory studies.

3. 2. Methods

Field work was carried out in 1983 and 1984. Samples were collected from seven sites which were chosen to include different degrees of elevation above the mean high water (MHW) level. For intertidal comparison, data obtained earlier are included (GERDES & HOLTKAMP, 1980). Elevation of sampling sites was measured with a theodolite.

Methods of sampling (sediments, microbial mats and fauna), thin section preparation and measurements of salinity and physicochemical parameters (Eh, pH, temperature) were identical with methods already described for the Gavish Sabkha. Additionally, relief casts were prepared (REINECK, 1970), and pigment concentrations in the upper 10 mm of sediments were determined using a corer 10 mm in diameter. Sediment cores for nutrient concentration measurements were taken with a corer 2.5 cm in diameter and 12 cm long. Pigments were extracted and measured using the methane-hexane extraction method (STAL et al., 1984a). Standard methods were used for extractions of NH_4^+ and S^{2-} (PANG & NRIAGU, 1976; HÖPNER et al., 1979; DEV, 1984). Diversity (after SHANNON) and eveness (after PILOU) of faunal communities were determined.

3. 3. Locality and previous work

3. 3. 1. Recent sedimentological history

Climatic fluctuations in the Quarternary led repeatedly to glaciations (Elsterian, Saalian and Weichselian) accompanied by exposures of wide parts of the North Sea basin on the one hand and sea level rises within interglacial periods on the other hand. The present-day configuration of the southern coastal area (Fig. 31A) is the result of the Holocene rise, which is by no means completed today (STREIF & KÖSTER, 1978). Longshore bars partly above and partly below the low water line and the formation of barriers at MHW-levels may have been the initial steps of the barrier-island formation, built up by combined forces of wave action and tide currents. It is highly probable that initially microbial films and later mats interfered with stationary approaches of sand bodies at MHW-levels. That microbial colonization and sediment immobilization immediately takes place with the emergence of sand bodies above MHW has been observed at stationary sand banks presently forming in open tidal flats between the Weser and Elbe estuary (REINECK & GERDES, 1984) and is also described in the following sections.

3. 3. 2. General setting of Mellum Island and study area

Within the chain of barrier islands which fringes the tidal flats along the southern and eastern North Sea coast (Fig. 31A), Mellum and some other smaller islands represent younger stages of barrier island development, called "bank islands" (GÖHREN, 1975). Bank islands occur mainly in the estuary embayment between the Jade Bay and the Eiderstaedt peninsula and typically represent mesotidal and macrotidal environments. According to DAVIES (1980), macrotidal environments have a tidal range of over 4 m at spring tide, while the range of mesotidal environments is somewhat between 2 m and 4 m at springs. With a spring tide of 3.40 m, Mellum Island is already close to a macrotidal environment. It forms the most westerly bank island of the estuary embayment of the "German Bight", where tidal flats extend widely in south-north direction as a result of larger tidal channels cutting through the tidal flats on either side of the islands (Fig. 31B).

For the purpose of this paper, the terms intertidal and supratidal are related to sediments which lie between or above the following tide levels: Intertidal: between mean low water at spring tide (MLWS) and mean high water at normal tide (MHW); Lower supratidal: between MHW and mean high water at spring tide (MHWS); Upper supratidal: above MHWS.

A characteristic feature of Mellum Island is the strong convexity of the front side, ending in two hooks recurved landwards with spits longitudinal to tidal inlets (Fig. 31B). This configuration protects a centered, slightly cliffed upper supratidal marsh and wide lower supratidal and intertidal sand flats.

The extension of the intertidal zone from MLWS to MHW is in the range of 1000 to 1200 m. It has a level difference between -1.80 m (MLWS) and +1.30 m (MHW), related to the Mean Sea Level which crosses the intertidal zone in the lower part. Microbial mats appear at the intertidal-supratidal zone boundary which parallels the western hook. Their distribution reaches upward to the MHWS-level and merges horizontally with the centered high salt marsh of the island (Fig. 31B). The distance between MHW and MHWS (lower and upper boundary of mat formation) is 400 to 700 m. A maximum elevation of 40 cm over this distance points to the extremely gentle angle of the lower supratidal zone.

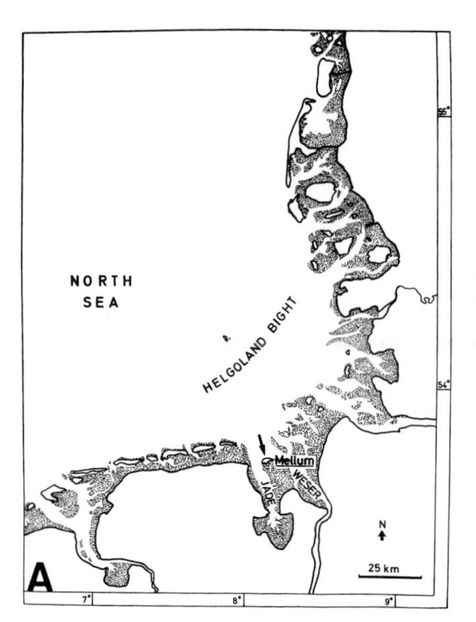

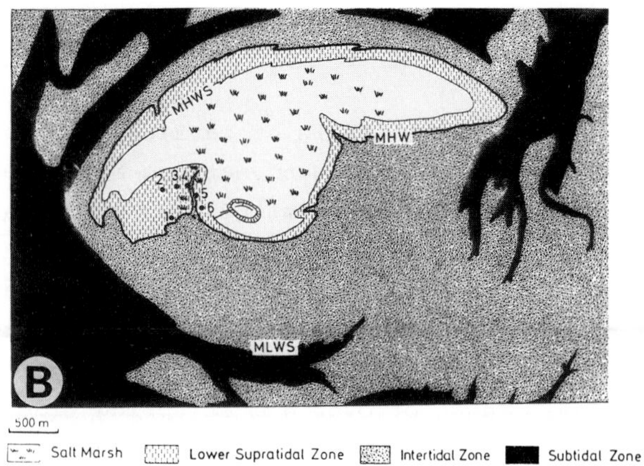

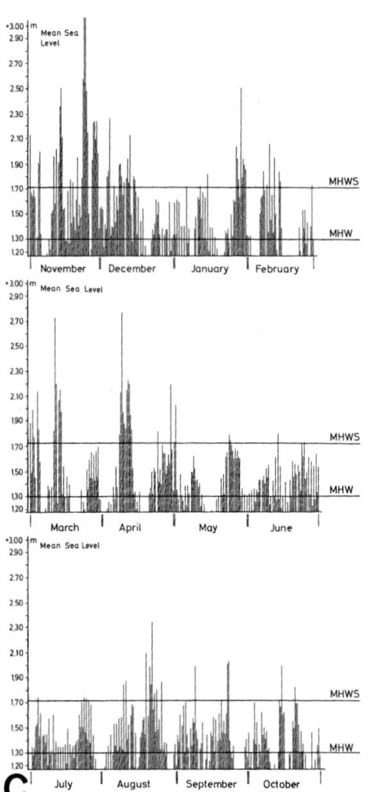

Fig. 31. General setting of Mellum Island.

A) Location of the island between the Jade Bay and the Weser river.

B) Study area, situated at the backside of the western hook, between the mean high water line (MHW) and the mean high water line at spring tide (MHWS). Numbers refer to sampling sites.

C) Ranges of high tides above MHW, n = 706 tides from November 1981 to October 1982. MHW = lower boundary of the versicolored flats, MHWS = upper boundary. In the annual cycle, areas between these lines are flooded from 70 % (MHW) to 20 % (MHWS). Extreme high water at spring tide (EHWS) occurs mainly in winter and spring. During summer, periods of exposure are more extended. (A to C with permission from J. Sediment. Petrol., 55, 265-278, 1985).

3. 3. 3. Previous work

The number of reports on biolaminated deposits in siliciclastic environments may be comparatively low if one regards the wealth of literature on carbonate stromatolites and microbial mats in tropical environments. However, biolaminations in tidal flats of higher latitudes have already attracted attention in preceding centuries and inspired taxonomic and ecologic studies on "conferves", (MÜLLER, 1777), later re-named "blue-green algae" and then cyanobacteria. At the beginning of the 19th Century, a group of Scandinavian natural scientists, including HOFMANN BANG, LYNGBY, OERSTEDT and ROSENBERG, dedicated their observations to microbial mats in lower supratidal settings of the Baltic Sea and emphasized the "island-forming capacity" of microorganisms, hidden from view (OERSTEDT, 1841; see also KRUMBEIN, 1986b). About a hundred years later, SCHULZ (1936), SCHULZ & MEYER (1940) and HOFFMANN (1942, 1949) studied the multiple interactive pathways of cyanobacteria and associated microbes in their sedimentary environment during visits to Amrum, a North Frisian island. The colorful laminations inspired SCHULZ (1936) to create the term "Farbstreifen-Sandwatt" (versicolored quartz-sandy tidal flats).

Nitrogen fixing by non-heterocystous cyanobacteria was intensively studied in siliciclastic tidal flats (STAL & KRUMBEIN, 1981; 1985; STAL & VAN GEMERDEN, 1984; STAL et al., 1984a, b c, 1985). Several examples of depositional records were given (EVANS, 1965; DAVIS, 1966, 1968; SCHWARZ et al., 1975; CAMERON et al., 1985; GERDES et al., 1985b, c; REINECK & GERDES, 1984). The aspect of coastal bio-engineering was emphasized by FÜHRBÖTER et al. (1981, 1983) and MANZENRIEDER (1984), who noted current velocities of more than 12 times the norm needed for grain movement which were made immobile through microbial colonization.

3. 4. The physical environment of mat formation

3. 4. 1. Climate

The southern coast of the North Sea has a temperate, humid climate. Low pressures, westerly winds, and high rainfall are typical. Seventy percent of the time, winds come from the southwest, or northwest with speeds of 6 to 15 m/s (EISMA, 1980). Precipitation averages 700 to

750 mm per year. During winter, the relative humidity is 90 - 95 %, while in the summer it is 75 - 80 %. Mean annual air temperatures are 4 °C in winter and 16 °C in summer. During winter, drift ice accumulation on the tidal flats is common (REINECK, 1976).

3. 4. 2. Flooding frequency

On a yearly average, 70 % of all high water reaches or crosses the MHW level, but only 20 % reaches or crosses the MHWS level (Fig. 31C). During the summer, spring tides mainly cross the lower supratidal slope only, while the upper supratidal zone remains subaerially exposed. Extreme high water at spring tide (EHWS) occur mostly between November and April and crosses the entire supratidal zone (Fig. 31C). Wind also influences the water level. Strong northwesterly winds raise it up to several meters above normal high-water level, resulting in the inundation of the supratidal area. On the other hand, winds from the east or south lower the water level for several dm, so that even high water at spring tide does not reach or cross the supratidal zone.

3. 4. 3. Salinity

Salinity in the high tide flats is a rather fluctuating factor. During rainfall at low tides, a minimum salinity of about 20 ‰ was measured in surface waters (the general range of the overall salinity of flood waters lies between 30 and 32 ‰). After a longer summertime period of insolation, a maximum salinity of up to 40 ‰ was found in interstitial water. SCHULZ & MEYER (1940) observed an even more drastic deviation from the normal seawater salinity. The authors measured 9 ‰ after a heavy rainfall and a salinity increase of up to 46 ‰ after a period of intense insolation. However, as already described for the Gavish Sabkha and the Solar Lake, an oscillating salinity regime does not control the mat formation in general, though it may probably select for special microbial taxa.

3. 4. 4. Moisture

The sedimentary surfaces of the lower supratidal zone remain moist even after a set of several days of subaerial exposure which is common in the summer period (see Fig. 31C). The surface-provided moisture at low tides is due to the continuous upward movement of groundwater, whose migration is supported by the fine-grained sandy texture (SCHULZ, 1936; HOFFMANN, 1942). Interstitial water contents obtained by dry weight determinations show values comparable to contents of intertidal sediments (Table 8; see also GERDES & HOLTKAMP, 1980).

TABLE 8. Weight percentage of interstitial fluids obtained after sectioning of sediment cores into 1 cm thick slices, drying of the sediment portions and calculation of the water loss. Sediment cores were taken at sites of equate elevation (MHW +30 cm) of the lower supratidal slope.

Sediment layers below surface	*Oscillatoria* variation	*Microcoleus* variation (shallows)
0-1 cm	18.7 wt.%	25.8 wt.%
1-2 cm	18.2 "	22.8 "
2-3 cm	15.0 "	22.4 "
3-4 cm	18.6 "	19.7 "
4-5 cm	16.7 "	17.3 "

3. 4. 5. Morphological unconformities

Two different sub-environments can be distinguished within the lower supratidal zone. The one, at the less protected seaward part of the western hook, has a relatively smooth surface topography without internal relief. The other is a more protected area at the junction between the hook and the centered salt marsh of the island. This area has a considerable amount of internal relief because of a flood and ebb channel system which crosses the junction and merges with the intertidal-supratidal zone boundary (Fig. 32A). The shallow and nearly filled flood channels are typical geomorphic elements of backshore environments. JACOBSEN (1980) used the term "landpriel shallows" to characterize such relief nonconformities of sandy coasts. The ebb channels drain the salt marsh hinterland. Both channel types are bordered by parallel-sided flood plains. The plains are flooded at MHWS and are subaerially exposed at MHW while surface water remains in the channels even at low tide. This results from ebb drainage on the one hand and

from small front bars on the other hand which build up in the vicinity of the MHW-level from inward-moving sand. Dense stands of halophytes characterize the higher-lying plains (Fig. 32A).

Fig. 32. Area and development of the versicolored tidal flats.
A) View of the area of the *Microcoleus* variation: Ebb channels border flood plains with stands of halophytes. Microbial mats interacting with low-rate sedimentation account for the elevation of the flood plains and make sediments rich in nutrients thus facilitating plant growth.
B) Initially, *Oscillatoria limosa* starts with the mat formation, colonizing and agglutinating sand grains. Scale is 200 μm.
C) Mature stage of the versicolored tidal flats showing a vertical zonation from top to bottom of (1) a cyanobacterial mat (usually covered with a thin sand layer), (2) anoxygenic photobacteria (purple bacteria), (3) chemoorganotrophic bacteria (sulfate-reducers). (Modified after GERDES et al., 1982).
D) Mat surface (*Microcoleus*) showing the densely entangled meshwork of filaments overgrowing the sandy sediment. Scale is 250 μm.
E) Filamentous cyanobacteria (*Oscillatoria limosa*) colonize the much thicker filament of a macroalgae (Enteromorpha sp.). Scale is 50 μm (after GERDES & KRUMBEIN, 1986).
F) Forams (undetermined species) are inhabitants of the mat surfaces. Scale is 200 μm.

3. 5. Sub-environments and facies

3. 5. 1. Local dominance of mat-producing species

Frame builders of the Mellum mats are cyanobacteria, including genera already mentioned for the Gavish Sabkha and the Solar Lake (*Microcoleus*, *Oscillatoria*, *Spirulina*, *Gloeothece*, *Synechococcus*). Two species of filamentous cyanobacteria share in the domination of the lower supratidal zone: *Oscillatoria limosa* and *Microcoleus chthonoplastes*. *O. limosa*, a very motile filamentous cyanobacterium, is dominant within the less protected and less vegetated seaward part of the lower supratidal zone where wind-drifted sediments settle at a high rate. The species is capable of a rapid gliding motility which aids these organisms to move fast, away from the growth site in response to the sedimentation events. The transitional part of the lower supratidal zone (landpriel shallows) merging into the centered saltmarsh is dominated by *Microcoleus chthonoplastes*. Specific slow sedimentation conditions characterize especially the flood plains where dense stands of halophytes (*Salicornia* sp., *Spartina anglica*, *Limonium vulgare*) occur (Fig. 32A). These conditions provide sufficient shelter for the ensheathed *Microcoleus* bundles to form mats. As already mentioned in the description of the Gavish Sabkha mats, the peculiarity of filaments of *Microcoleus chthonoplastes* is to move horizontally back and forth and to leave their sheaths only in crisis conditions. The geobiological importance of *M. chthonoplastes* in the Gavish Sabkha and the Solar Lake (formation of L_h-laminae, high productivity rates and sheaths which are relatively recalcitrant and survive decomposition) was also emphasized. The dominance of this species within the protected part of the lower supratidal zone of Mellum Island results in biolaminites which are much more vertically extended than those of *Oscillatoria limosa*. *O. limosa* seems to represent a typical opportunist which colonizes predominantly freshly deposited sediments (e. g. on the flats off the shallows and in the vicinity of the MHW-level) while *M. chthonoplastes* seems to benefit from the sediment stabilizing activity of *O. limosa* and low-rate sedimentation.

According to the change of dominance and abundance of these two species, we denote the sub-environment of the outer slope *Oscillatoria* variation, the transition to the centered marsh *Microcoleus* variation.

3. 5. 2. Stratification of living top mats

Osciallatoria-Variation: Sections through the living top mat usually show a vertically structured set of three to four layers:

1. A yellow-brown top layer, composed of sand. Thickness ranges from 1 to 5 mm. The layer is colonized by diatoms (the genera *Navicula* and *Diploneis* are dominant).
2. A greenish layer underneath, composed of filamentous cyanobacteria. Thickness ranges from 0.5 to 1 mm. *Oscillatoria limosa* is the dominant species (Fig. 32B). Also present are bundles of *Microcoleus chthonoplastes*, other filamentous forms of the genera *Oscillatoria*, *Spirulina*, *Lyngbya*, *Phormidium*, and clusters of unicellular cyanobacteria (*Gloeocapsa*, *Gloeothece*, *Synechocystis*, *Merismopedia*).
3. Where the underside of layer 2 is aerated, a thin (usually not more than 0.2 mm thick) red orange horizon forms which is rich in iron hydroxide.
4. A layer black with iron sulfide ranging in thickness between 1 and 5 mm follows subsequently or in place of the iron hydroxide.

Microcoleus variation: A well-defined change in associations of microbes characterizes this variation. The vertically structured system of the *Microcoleus*-dominated microbial mat (Fig. 32C) includes four to five layers:

1. The top mat is usually 50 - 100 µm thick and monospecifically composed of bundled sheaths of *Microcoleus chthonoplastes* which are horizontal in orientation. This top mat contains a minor number of quartz grains (Fig. 32D);
2. Below is a sandy layer, 600 to 700 µm thick, completely green colored by dominance of filamentous and unicellular cyanobacteria. The organisms colonize sand grains and interstices (Fig. 32C). Filamentous organisms (*Oscillatoria, Lyngbya, Phormidium*) are predominantly vertical in orientation. The representatives of unicellular cyanobacteria are identical with those of the *Oscillatoria* variation of the outer sand flats (*Gloeocapsa, Gloeothece, Synechocystis, Merismopedia*).
3. The third layer is pink from the dominance of purple sulfur bacteria (photosynthetic anoxygenic bacteria, mainly *Chromatium vinosum* and other non-identified species). Sand grains and grain-supported interstices are densely colonized by the microorganisms. The thickness of this layer ranges between 500 and 600 µm.
4. The colorful vertical zonation merges into a fourth layer which is black from iron sulfide and reaches down to several cm.

The top mat is sometimes covered by a thin layer of yellowish white, cleanly washed sand. Diatoms are not as frequently encountered in this layer as in the *Oscillatoria* variation.

The *Microcoleus* variation represents the mature stage while the *Oscillatoria* variation may indicate the pioneer stage of mat development. The accumulation of organically bound carbon, nitrogen, sulfur and other compounds within the nutrient-poor quartz-sand facilitates the enrichment of other metabolic types. Initially these are aerobic chemoorganotrophic bacteria which degrade the primarily monolayered mat. With the constant rise of organic matter, O_2 becomes depleted, which gives rise to the development of anaerobic chemoorganotrophic bacteria (e. g. *Desulfovibrio* sp.). Finally, the versicolored, vertically structured system, already described for the Gavish Sabkha and the Solar Lake, develops consisting of a great variety of metabolic types. The system benefits from a setting which is protected against high sedimentation rates, wave attack and tidal currents.

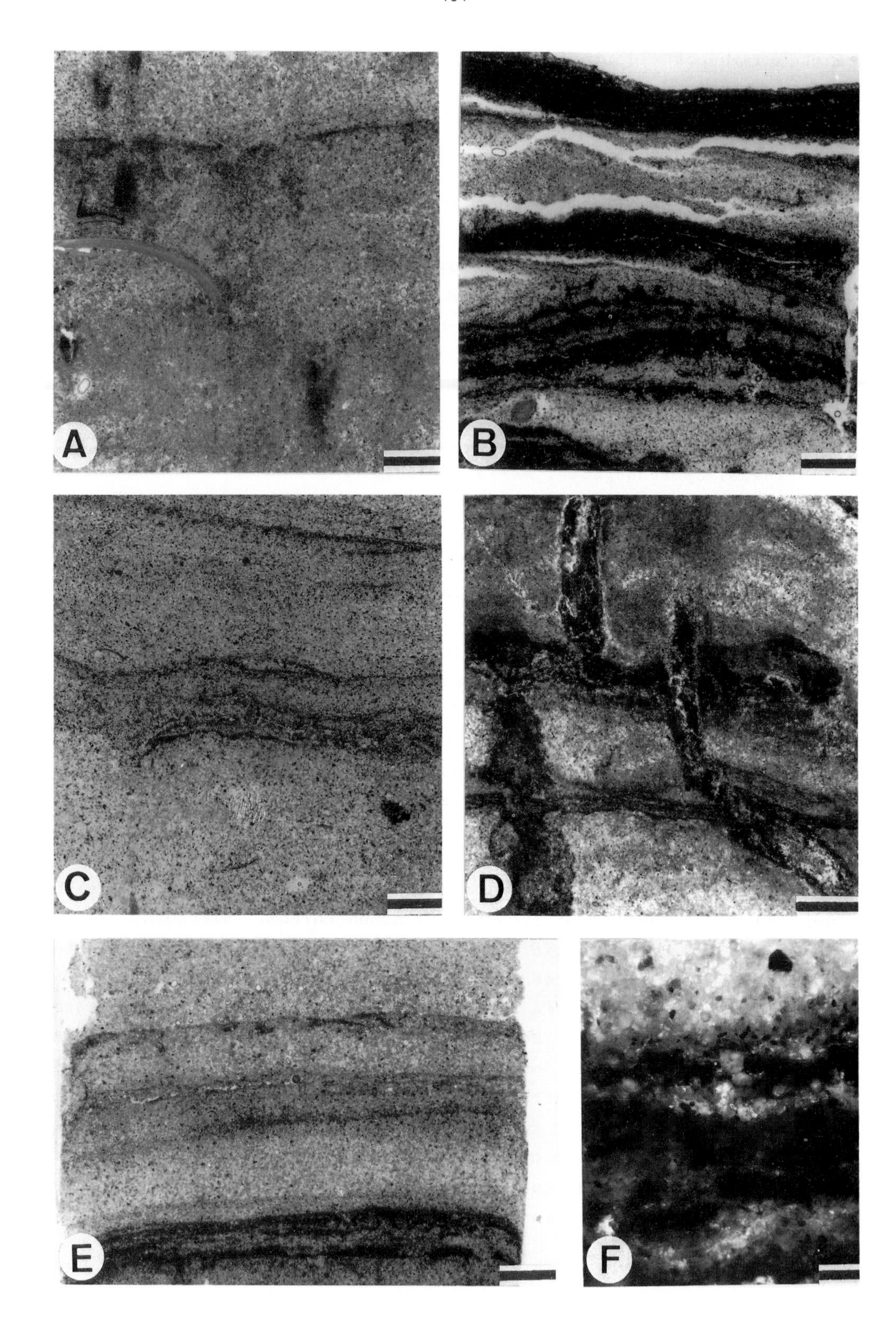
A
B
C
D
E
F

Fig. 33. Internal sedimentary structures (thin sections from cores). A - E: Photographs by H.-E. REINECK.
A) Thin laminae characteristic of the *Oscillatoria* variation. Dark vertical lines are tubes of polychaets. A shell fragment is visible below the mat. Scale is 5 mm.
B) Multilaminated sequences within the *Microcoleus* variation, interlayered with light quartz-sand deposits. Scale is 5 mm.
C) A thin mat has coated a rippled surface. Scale is 5 mm.
D) Root shafts of *Spartina anglica* piercing through buried microbial mats. Note filling of root shafts at top and bottom right with sand. Scale is 1 cm.
E) Thinner laminae on top of a multilaminated mat sequence consist of heavy mineral grains indicating the former presence of mats. Even without preservation, the captured heavy mineral grains may indicate the former presence of a mat. Scale is 5 mm.
F) Close-up of the placer of heavy minerals enriched on top of a buried mat. Scale is 1 mm. (Figs. B to E with permission from J. Sediment. Petrol., 55, 265-278, 1985).

3. 5. 3. Internal sedimentary structures

Thin sections show that the sediments are almost invariably sandwiched between layers of cleanly washed sand and organic carbon-rich microbial mats which proliferate during intervals between sedimentation.

The grain-size distribution of the sandy layers is almost identical with intertidal sediments (GERDES & HOLTKAMP, 1980). The main fraction is made up of fine-grained sand with a medium grain size of 100 µm (= 3.3 Phi). A fraction of medium-grained sand is generally admixed averaging 20 to 25 wt.%, while silt and clay is not more than 3 wt.%. The lithologic conformity of sediments shows that the facies modifications are mainly controlled by inundation frequencies to which microbial and faunal assemblages (see chapter 3. 6.) correspond.

The buried mats appear as compacted, sharply projecting interlayered bands. In the *Oscillatoria* variation, laminae are usually thin (1 to 2 mm; Fig. 33A), while in the *Microcoleus* variation they appear extremely thick (up to 10 mm; Fig. 33B). Light layers between the mats indicate sedimentation units of quartz-sand which range in thickness between 1 mm (*Microcoleus* variation) and several cm (*Oscillatoria* variation). The biolaminites are mainly horizontal in orientation and predominantly planar, due to the predominance of parallel-laminated sand. It is, however, possible that a mat covers a small-scale wave ripple (Fig. 33C). Root shafts of halophytes appear with increasing height above MHW. The dense stands of halophytes within the *Microcoleus* variation (Fig. 32A) are spectacularly recorded by the root systems piercing through the buried microbial mats (Fig. 33D). Deposits of orange-red $Fe(OH)_3$ around the roots and filled root shafts occur (Fig. 33D). A spectacular arrangement of heavy mineral grains following the morphology of a buried mat is often visible (Fig. 33E and F). Heavy mineral is concentrated in sediments around Mellum Island with bulk volumes ranging between 0.2 to 3.5 wt.%. Epidot is the most abundant mineral (LITTLE-GADOW, 1978). The concentration of grains on top of supratidal mats may indicate wind activity and deflation (TRUSHEIM, 1935; GADOW & REINECK, 1969; REINECK & SINGH, 1980).

3. 5. 4. Standing crops and biogeochemistry

Standing crops. Pigment concentrations confirm the different complexity of both mat types described (Fig. 34A). In the course of the summer the concentrations increase in both subenvironments. Values of chlorophyll a contents in the *Oscillatoria* variation (ranging between 50 and 170 mg x m^{-2}) are comparable to those usually found in upper-intertidal flats (VAN DEN HOEK et al., 1979). The concentrations in the *Microcoleus* variations are up to four times higher, and bacteriochlorophyll a is maintained only in samples of this variation. The values of the Microcoleus mat are almost comparable to those of the Gavish Sabkha and the Solar Lake.

The drastic increase of pigment contents in October (Fig. 34A) can be traced back to a bloom of macroalgae (*Enteromorpha* sp.), initiated by increasing moisture supply (rain) in fall. Thus, the microbial biomass can best be estimated when drier weather favors the drought-resistant cyanobacteria and prevents growth of the macroalgae (e. g. in August, Fig. 34A). Though their growth is ephemeral, the macroalgae effectively contribute to the biomass production and also to the complexity of the *Microcoleus* mat, since the cyanobacteria tend to intertwine with the much thicker filaments of the algae (Fig. 32E).

The macroalgae (mainly *Entermorpha* sp.) are particularly abundant in the *Microcoleus* variation while almost absent in the *Oscillatoria* variation. *Enteromorpha* species are able to use ammonium directly without further break-down to nitrate (RANWELL, 1972). Hence, the

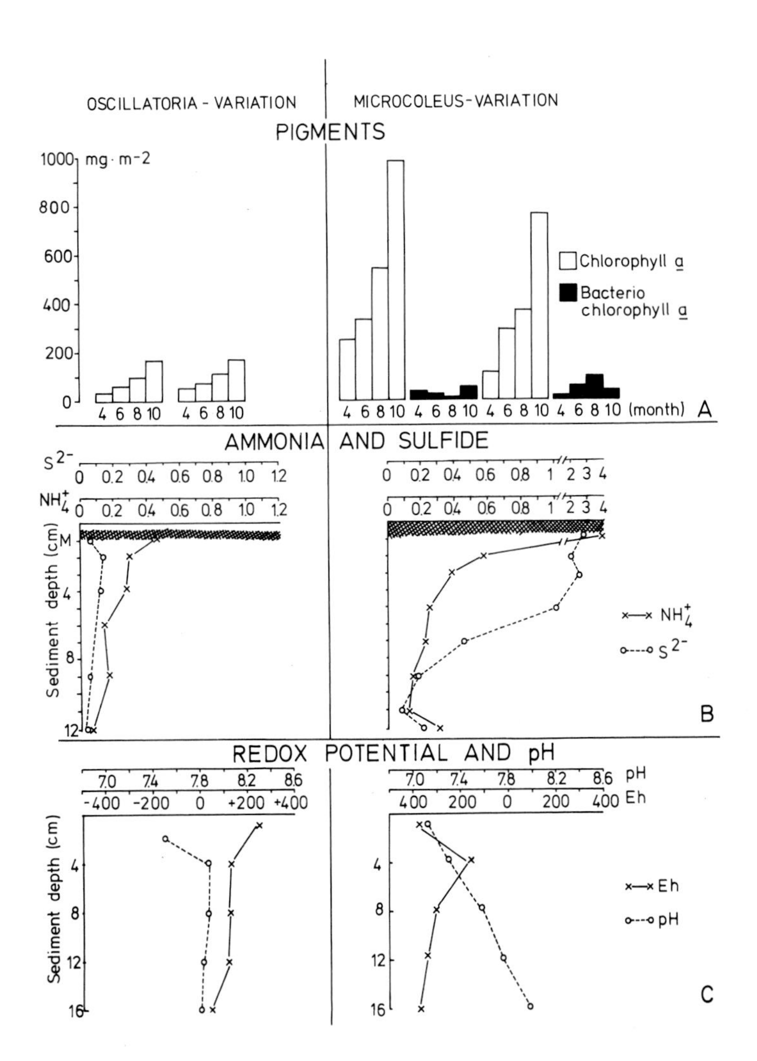

Fig. 34. Substrate parameters in the two variations of the versicolored tidal flats. Left: *Oscillatoria* variation (unprotected sites of the lower supratidal slope), right: *Microcoleus* variation (protected landpriel shallows).

A) Chlorophyll a and bacteriochlorophyll a in surface mats in the course of the summer 1983. The greater complexity of *Microcoleus* mats (right) can be read from (1) chlorophyll a concentrations absolutely higher than of the *Oscillatoria* mats (left), (2) a more drastic summerly increase of chlorophyll a and (3) the occurrence of bacteriochlorophyll a indicating the presence of anoxygenic photosynthetic bacteria (e. g. purple bacteria).

B) Distribution of ammonia and sulfide vs sediment depth. Both rates are increased in near-surface sediments, in particular in the *Microcoleus* variation and less in the *Oscillatoria* variation. The profile of the *Microcoleus* variation indicates a site with stands of macroalgae whose growth is favored by the especially high ammonia values (compare Fig. 32E).

C) Redox potentials and pH vs sediment depth showing positive Eh-values and slightly higher pH in the *Oscillatoria* variation, negative Eh-values and lower pH in the *Microcoleus* variation.

selected growth of these species may suggest that ammonium production is high within the sediments of the *Microcoleus* variation.

Ammonia and sulfide concentrations. In October 1984 we took sediment cores from both subenvironments to study the vertical distribution of NH_4^+ and S^{2-}. Both rates were largely increased in the *Microcoleus* variation and far less important in the *Oscillatoria* variation. The concentrations were particularly high in the near-surface parts and decreased more or less rapidly with depth (Fig. 34B).

This is consistent with the presence of freshly produced organic matter which is more easily degradable by chemoorganotrophic bacteria than older and more resistant material (BLACKBURN, 1983). The near-surface pool of ammonium explains why macroalgae capable of using the reduced intermediate product of N-mineralization are attracted to colonize the mats. On the other hand, the enrichment of sulfide just below the surface which is associated with the bacterial break-down of dead organic matter provides a beneficial situation for anoxygenic phototrophs which require light. Thus, the sandy sediments immediately below the aerated surface mat maintain a large and active population of purple sulfur bacteria as well as green and colorless sulfur bacteria (compare Fig. 32C). The growth of the sulfur bacteria is inhibited as soon as oxygen penetrates below the surface mat. This is mainly realized in the *Oscillatoria* variation where (1) physical processes such as wave energy and (2) the less condensed fabric of the mat support the penetration of oxygen from above into the system. Furthermore, the *Oscillatoria* mat is characterized by a lower productivity rate of biomass and thus by a lower amount of dead organic matter to be broken down by bacteria to ammonium. This may explain why macroalgae as well as larger and more active anoxygenic phototrophs are lacking there.

Reduction-oxidation potentials and pH. In the *Oscillatoria* variation, Eh-values were mainly positive (Fig. 34C) and pH in interstitial water below the mat slightly alkaline typical of bicarbonate buffered seawater. The values measured in this subenvironment are comparable to those at the Gavish Sabkha sandlobes and gullies (Fig. 12).

The *Microcoleus* variation is characterized by anoxic conditions, which indicate that more oxygen is respired than is provided from above. The high amounts of organic matter, as well as the effective diffusion barriers provided by the mats, allow the enrichment of chemoorganotrophic bacteria which are able to use sulfate and sulfur as electron acceptors (JORGENSEN, 1977).

Though condensed layers exist in the *Microcoleus* variation in the form of buried mats, interstitial water of higher contents as in the *Oscillatoria* variation were usually found (Table 8). This may be due to the lateral migration of water from the creeks and shallows into the adjacent flood plain deposits where mats predominantly form. The condensed layers of buried mats in turn support stagnancy of interstitial water which allows the enrichment of chemical reductants.

3. 6. Fauna and ichnofabrics

3. 6. 1. Mixed marine-terrestrial composition

In the lower supratidal zone, 40 species of both macrobenthic and meiobenthic invertebrates were identified (Table 9). The macrobenthic invertebrates include 12 species of marine and 5 species of terrestrial provenance.

Marine macrobenthic invertebrates. Representatives of the marine intertidal fauna including small polychaetes, oligochaetes, amphipodes, gastropodes and lamellibranchs attain highest densities (Table 10). The highest density (about 56,000 individuals beyond 1 m^2) was found at the intertidal-supratidal zone boundary. Towards MHWS, the number of species and individuals of intertidal invertebrates decreases. Five species still remain (*Pygospio elegans*, *Corophium arenarium*, *Hydrobia ulvae*, *Nereis diversicolor* and *Lumbricillus lineatus*). The faunal assemblage composed of these five species attains high densities up to MHWS (a maximum 30,000 individuals and more beyond 1 m^2 was encountered). This density is amazing if we consider that the MHWS-level is reached or crossed by only 20 % of high waters in the yearly average (compare Fig. 31C).

Terrestrial invertebrates. The five terrestrial species found include three coleopteran and two dipteran species (Table 9). Their occurrence is rather patchy and restricted to sites close to the MHWS-level. At some places, however, they occur side-by-side with the marine fauna. Mainly beetles contribute to the lebensspuren spectrum. Two

TABLE 9. Benthos fauna of the versicolored tidal flats, Mellum Island, lower supratidal zone. (*: restricted by increasing height above MHW)

MAJOR TAXA		SPECIES
		1. MACROBENTHOS
MARINE:		
CRUSTACEA	AMPHIPODA	*Corophium arenarium*
		Bathyporeia sp.*
ANNELIDA	CLITELLATA	*Lumbricillus lineatus*
		*Lumbrificoides benedeni**
	POLYCHAETA	*Pygospio elegans*
		Nereis diversicolor
		*Eteone longa**
		*Capitella capitata**
		*Heteromastus filiformis**
		*Fabricia sabella**
MOLLUSCA	BIVALVIA	*Macoma baltica* *
	GASTROPODA	*Hydrobia ulvae*
TERRESTRIAL:		
INSECTA	COLEOPTERA	*Bledius spectabilis* ssp. *frisius*
		Bledius subniger
		Heterocerus flexuosus
	DIPTERA	*Scatella subguttata*
		Dolichopodidae sp.

species of the staphilinide genus *Bledius*, which was already emphasized for the Gavish Sabkha and the Solar Lake, are present, *B. subniger* and *B. spectabilis* v. *frisius*. Even the ecological zonation, the burrowing and feeding behavior of the Mellum species resembles those of the Gavish Sabkha species: *Bledius subniger* (Mellum) and *B. angustus* (Gavish Sabkha) are both of a smaller size which prefer oxic sandy deposits and feed preferentially on unicellular cyanobacteria and diatoms, while the *B. spectabilis* (Mellum) and *B. capra* (Gavish Sabkha) are both larger species which dominate the thickened and mainly anoxic biolaminites. On Mellum, these thickened, *Microcoleus*-dominated biolaminites also provide an adequate substrate for another burrowing beetle, *Heterocerus flexuosus*.

Meiofauna. Of the 23 species identified nematodes are the most diverse systematic group (17 species). Nematodes occur in especially high individual numbers in the thickened biolaminites of the *Microcoleus* variation. A vertical distribution occurs also on Mellum Island

Table 9 continued

2. MEIO- AND MICROZOOBENTHOS

CRUSTACEA	OSTRACODA	*Leptocythere baltica*
		Leptocythere lacertosa
		Elofsonia baltica
	COPEPODA	*Mesochra lilljeborgi*
		Heterolaophonte minuta
		Tachidius discipes
NEMATODES		*Enoplus brevis*
		Enoploides labiatus
		Viscosia sp.
		Adoncholaimus fuscus
		Oncholaimus paroxyuris
		Bathylaimus australis
		Tripyloides marinus
		Diplolaimelloides deconincki
		Daptonema sp.
		Theristes acer
		Chromadora nudicaptata
		Ascolaimus elongatus
		Sphaerolaimus gracilis
		Hypodentolaimus balticus
		Praeacanthonchus punctatus
		Halichoanolaimus robustus
		Desmodora communis
TURBELLARIA		
ROTATORIA		various species not determined
CILIATA		
FORAMINIFERA		

and thus confirms SCHULZ' observations (SCHULZ, 1936), although GERLACH's observation on a Farbstreifen-Sandwatt at the Danish coast (GERLACH, 1977) indicates that the green oxygenated layer of the cyanobacterial mats was favored by most nematodes. Very few were found in the underlying anoxic strata.

Ostracods (3 species) and copepods (3 species) mainly dominate the aerated mat surface and here outnumber all other meiobenthic organisms. Forams (representatives of the microfauna, Fig. 32F) also prefer to colonize the mat surfaces.

3. 6. 2. Trophic types

Calculations of dominance and diversity structures of trophic types present in Recent and fossil faunal assemblages are often used to interpret gradients in food resources and physical regimes. The main trophic categories used are (1) suspension feeders, (2) deposit feeders, (3) grazers, (4) predators.

Food particles held in suspension for solely suspension-feeding organisms on the one hand and sedimentated detritus for solely deposition-feeding organisms on the other hand are rarely found in the zone above MHW, while bacteria of both filamentous and unicellular shape, micro- and macroalgae are enriched. The dominance of grazers is thus to be expected. This is at least confirmed by terrestrial species, while none of the trophic categories mentioned above is found in isolation with the marine macrofauna. We thus introduce here alternative terms of (1) mixed feeders, (2) sandlicker and (3) bacteria feeder which characterize more expressively the peculiar situation of coexistence of microbial mats and marine benthic fauna in a high tide flat.

Mixed feeders. The tube-building polychaete *Pygospio elegans* characterizes a species which is able to switch from suspension feeding during ebb- and flood currents with higher velocities to deposition feeding during slower current velocities (GALLAGHER et al., 1983). We observed the feeding activity of animals of this species cultured in aquarium tanks with microbial mats: The animals came out of their tubes with half of the body length and collected diatoms, unicelluar cyanobacteria as well as filaments of *Oscillatoria limosa*. Sand-grains coated with microorganisms were also ingested. Even bundles of *Microcoleus chthonoplastes* enclosed by their very tough extracellular sheathes (Fig. 32D) were used. The animals swallowed one end of a sheath-enclosed bundle and ripped on the tough sheath material until enclosed filaments became liberated. These were ingested. The energy costs may be, however, very expensive for an animal which does not possess jaws to utilize *Microcoleus* as the major diet. This may be one of the reasons why *Pygospio elegans* is sparsely distributed in the *Microcoleus* variation.

The second mixed feeder occurring in both local variations is the burrowing polychaete *Nereis diversicolor*. The species is known to be a grazer, suspension- and depositional feeder and is also predacious. In contrast to *Pygospio elegans*, this species possesses heavy jaws which aid the feeding on tough substrates such as *Microcoleus* mats. Fecal pellet contents after a diet of microbial mats show a great variety of more or less empty cyanobacterial sheaths (Fig. 35).

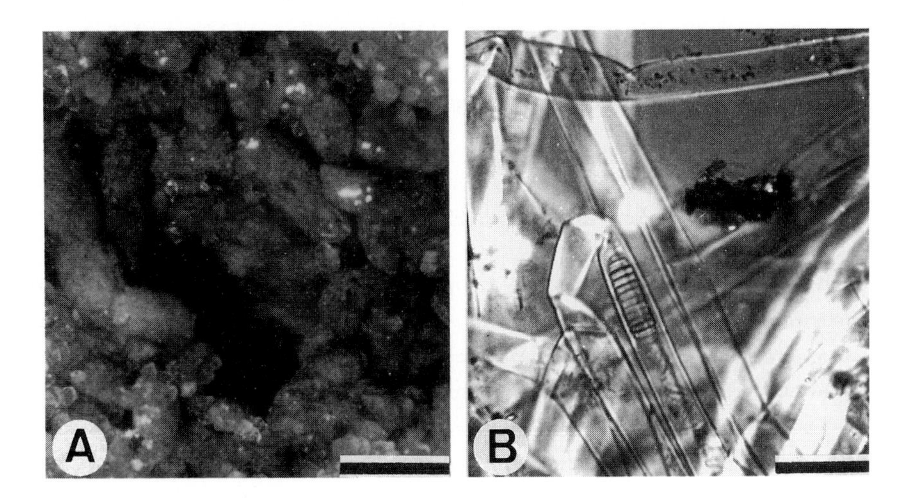

Fig. 35. Grazing on mats.
A) Entrance of a burrow of the polychaet *Nereis diversicolor* within microbial mats surrounded by fecal pellets. Scale is 1 mm.
B) Light microscopy of the fecal pellets showing empty sheaths and fragmentated cells of *Oscillatoria limosa*. Scale is 100 µm.

Thirdly, also the mud snail *Hydrobia ulvae* may be characterized as a versatile feeder which ingests small sediment particles, diatoms and unicellular cyanobacteria.

Sandlicker. The term was used by REMANE (1940). Two species belong to this specialized trophic type: the burrowing amphipode *Corophium arenarium* and the beetle *Bledius subniger*. Both are found in the high tide flats and feed on bacteria and microalgae which colonize the single quartz grain (Fig. 35C). The animals scrape away the coatings with their mandibles and deposit the cleaned quartz grains around the burrows on the surface.

Bacteria feeder. Many depositional feeders are not really feeding on deposited particles themselves but ingest the bacteria coatings around organic detritus. This is the case also with *Lumbricillus lineatus*, the dominant marine invertebrate of the *Microcoleus* variation. The benefit of this species are the numerous bacteria populations which are associated with the *Microcoleus chthonoplastes* mats.

Grazer. Both coleopteran species *Bledius spectabilis* and *Heterocerus flexuosus* browse preferentially on the *Microcoleus* mats. Both possess heavy mandibles to rip away the tough mat material.

3. 6. 3. Regional distribution of trophic types

The following correlation of trophic types and local variations in mat development is recognized (the terms arenophile and chthonophile relate to preferences of the species for either quartz-sand or soil- = Greek chthonos -like substrates; the term chthonophile was chosen here to emphasize the role of *M. chthonoplastes* in forming a soil-like substrate in an otherwise quartz-sandy depositional environment):

Arenophile species	Intermediate position	Chthonophile species
(*Oscillatoria* variation)		(*Microcoleus* variation)
Sandlicker:	Mixed feeder:	a)Bacteria feeder:
Corophium arenarium	*Pygospio elegans*	*Lumbricillus lineatus*
Bledius subniger	*Nereis diversicolor*	b) Grazer:
		Hydrobia ulvae,
		Bledius spectabilis
		Heterocerus flexuosus

3. 6. 4. Life habits and ichnofabrics

The major trace makers and traces in the biolaminated sediments are listed below. Their differentiation into internal vs. surface structures used follows HERTWECK (1978). The classification scheme of SEILACHER (1953), who classified traces according to ethologic-functional categories (dwelling, grazing etc.), is also adopted.

I. Trace-making marine invertebrates

Pygospio elegans (Polychaeta, Spionidae). Infaunal semi-sessile mixed feeder (suspension-/deposit-feeder, grazer). Distribution: Intertidal and lower supratidal flats (preference for stabilized sand).
Lebensspuren: a) Internal: Dwelling trace. Tube with 5 mm overall diameter, vertical in orientation, length ranging between 5 and 10 cm. Tube wall: Mucus-cemented sand grains. Often red-orange coa-

tings of iron-hydroxide due to irrigation activity of the animal. b) Surface: Pore-like hole, not significant in fine-grained sand.

Corophium arenarium (Crustacea, Amphipoda). Infaunal semi-sessile deposit feeder. Distribution: Intertidal and lower supratidal flats (preference for stabilized oxygenated sand).

Lebensspuren: a) Internal: Dwelling trace. Burrow with 5 to 10 mm overall diameter, vertical in orientation, length about 5 cm. Burrow wall: Mucus-agglutinated (non-cemented) sand grains. b) Surface: Feeding trace. Toothed-wheel-shaped scratch marks, 5 to 10 mm overall diameter (Fig. 36A).

Nereis diversicolor (Polychaeta). Infaunal semi-sessile mixed feeder (suspension-/deposit feeder, grazer, predator). Distribution: Subtidal, intertidal and lower supratidal, muddy and sandy substrates.

Lebensspuren: a) Internal: Dwelling trace. Burrow with 5 to 10 mm overall diameter, vertical in orientation, length ranging between 5 and 10 cm. Burrow walls: Non-solid mucus-coatings (burrows within anoxic layers show $Fe(OH)_3$-coatings due to irrigation of oxygenated water). b) Surface: Feeding trace. Mucus-agglutinated, surface-furrowing traces of a characteristic deer-antler-shape. This occurs because when emerging from its burrow to seek food the animal remains in contact with the burrow with the tip of its tail. In this way it constantly moves sidewards and back and forth within the same trace to the burrow (Fig. 36B).

Hydrobia ulvae (Gastropoda). Epifaunal mobile grazer/deposit-feeder. Distribution: Intertidal and lower supratidal zone, muddy and sandy substrates.

Lebensspuren: a) Internal: Resting trace. Non-significant deformational bioturbation structure, few mm below surface (the snails survive drought periods by burying themselves in the sediment (Fig. 36C). Burrowing does not occur at low tide when surface water persists (VADER, 1964). b) Surface: Browsing trace. Narrow furrows, bilobate on muddy surfaces or when diatoms and cyanobacteria form a water-saturated mush at the surface of sand flats (Fig. 36D).

II. Trace-making terrestrial invertebrates

Bledius subniger (Coleoptera, Staphilinidae). Endofaunal mobile grazer. Distribution: Lower and upper supratidal, sandy oxygenated substrates.

Lebensspuren: a) Internal: Dwelling and browsing traces. Burrows vertical in orientation, 3 to 5 mm in diameter, length 2 to 3 cm.

Fig. 36. Feeding, resting and dwelling traces.
A) Toothed-wheel-shaped scratch marks at the surface are feeding traces of the amphipod *Corophium arenarium*. Photograph by H.-E. REINECK.
B) Feeding traces of *Nereis diversicolor* furrow the surface which is viscose by microbial slime. Scale is 1 cm. Photograph by W. HÄNTZSCHEL.
C) Low-tide resting snails (*Hydrobia ulvae*) below the surface. Scale is 2 mm.
D) Elongated piles of excavation pellets of the salt beetle *Bledius subniger*, removed from its subsurficial dwelling and browsing tunnels.
E) Round piles of excavation pellets of *Bledius spectabilis*. D and E: Photograph by H.-E. REINECK.
F) Browsing traces of the beetle *Heterocerus flexuosus* cut through a mat. Scale is 5 mm.

Browsing trace: Two-sided furrows leading from the burrow entrance a few mm below surface, horizontal in orientation, 1 to 2 mm wide, up to 5 cm long (similar to *B. angustus*, see Gavish Sabkha, Fig. 20B). b) Surface: Excavation pellets. Elongated piles of excavation pellets which the beetle removes from its dwelling and browsing tunnel (Fig. 36E).

Bledius spectabilis v. *frisius* (Coleoptera, Staphilinidae). Endofaunal semi-mobile grazer. Distribution: Mainly upper and lower supratidal salt marshes.

Lebensspuren: a) Internal: Dwelling trace (browsing traces not observed). Burrow with a diagonal top shaft (bottle neck) and a vertical shaft with breeding chambers branching off of the main shaft (similar to *B. capra*, see Gavish Sabkha, Fig. 18). b) Surface: Excavation pellets. Round piles of excavation pellets removed from dwelling burrows (Fig. 36F).

Heterocerus flexuosus (Coleoptera, Carabiidae). Endofaunal mobile grazer. Distribution: Mainly upper and lower supratidal salt marshes.

Lebensspuren (only internal): Dwelling and browsing traces. Dwelling: Burrow horizontal in orientation, several cm long. Burrows found on Mellum ran on top of a buried *Microcoleus* mat: Vertical section shows a fenestra-like elongated cavity with a flat floor and a convex thin roof layer composed of microbe-bound quartz sand which allows a coherent roof architecture (GERDES et al., 1985c). Browsing: Burrow horizontal in orientation, similar to the description of dwelling trace, but commonly deeper below surface. Vertical section shows cavities sometimes cutting through the horizontally orientated *Microcoleus* mat (Fig. 36G).

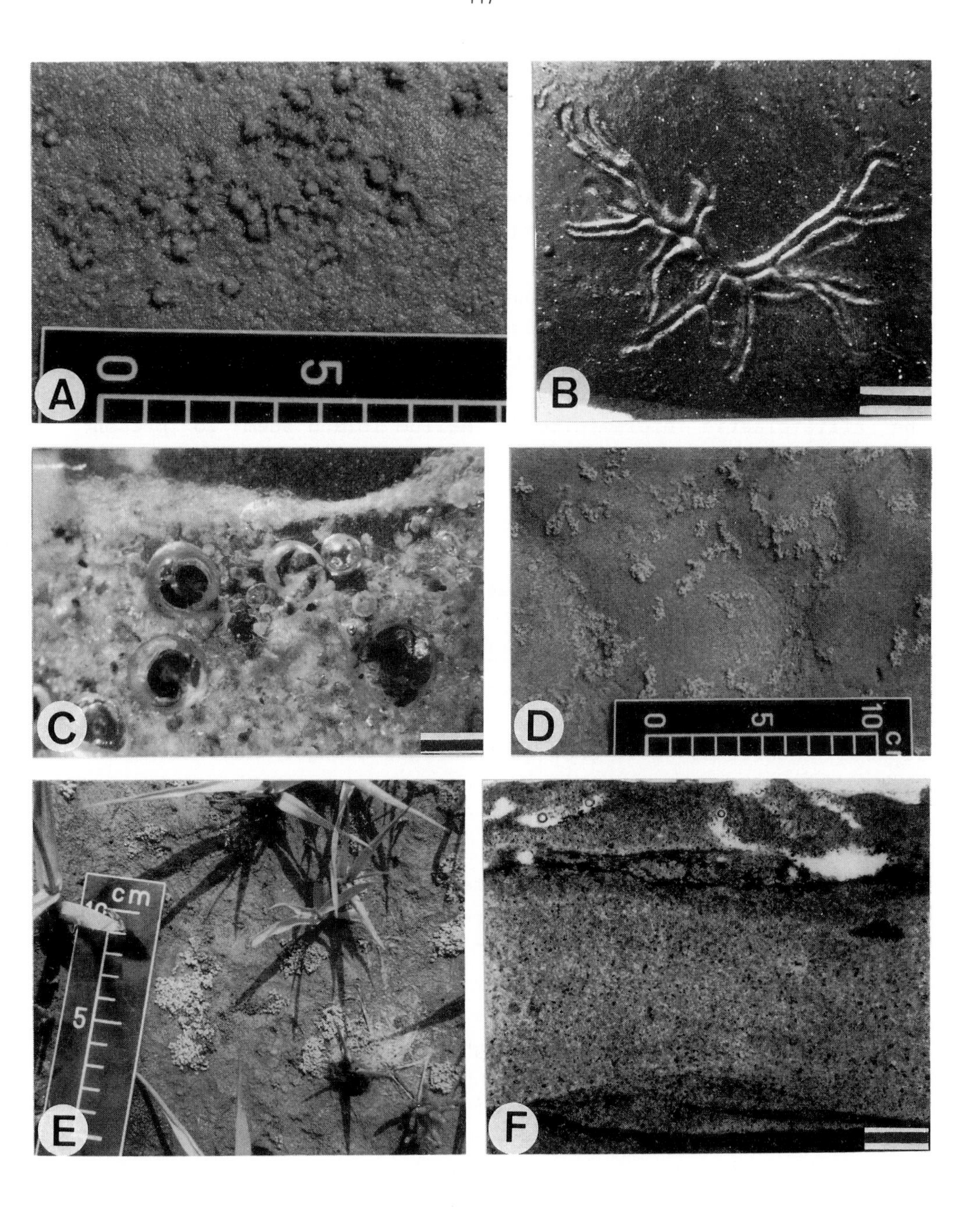

Examples where lithology influences the visibility of lebensspuren are numerous. It has thus to be emphasized that the ichnological patterns of species listed before refer to the sandy substrate in which the species at Mellum Island live, as most of them are also able to live in muddy tidal flats. For example, the polychaete *Heteromastus filiformis* is an important tracemaker in muddy substrates, but in the sand flats of Mellum Island, traces of this species are not significant and thus are neglected here.

3. 7. Dominance change and its importance for bioturbation grades and patterns

Sparse trace records are usual for stromatolites. Harshness of physical conditions is often suggested as a reason and is attested to be necessary for the exclusion of disturbances by grazing and bioturbation (AWRAMIK, 1981; GARRETT, 1970; GEBELEIN, 1976). In this respect the siliciclastic biolaminites of Mellum Island seem to present an unusual stromatolite environment in that they locally harbor an abundant marine fauna. However, the density of a faunal assemblage is not always responsible for the grade of churning and stirring of a sediment. Moreover, the question of species dominance and abundance in a given setting should also be addressed (CARNEY, 1981).

We will show by combined studies of faunal compositions (Table 10) and bioturbation structures preserved in relief casts (Fig. 37) that the faunal assemblages in the lower supratidal zone compared with those at the intertidal base show well-defined changes in dominance and abundance and that these changes account for the type and the grade of bioturbation within the mats.

According to SCHÄFER (1972) the terms "deformative" and "figurative" are used to characterize the differences in the type and the grade of bioturbation. The term "deformative" describes completely reworked beds in which no preexisting sedimentary structures can be seen (REINECK, 1967). Figurative bioturbation is maintained by the occurrence of burrows and tubes piercing through several sets of strata without destroying them completely.

3. 7. 1. Effects of increasing elevation

a) Lower intertidal base

The dominant type is the cockle *Cerastoderma edule*, and it is associated with several species of higher density: *Scoloplos armiger*, *Urothöe* sp., *Arenicola marina*.

TABLE 10. Distribution, abundance, Shannon's diversity and dominance calculations of macrofauna at different niveau levels (Abundance x m^2). Bioturbation structures corresponding with faunal composition and abundance are shown in Figs. 37A, B, D and F.

Species	Individuals m^{-2}			
	Lower Intertidal zone	Upper Intertidal zone	Lower supratidal zone Oscillatoria-variations	Lower supratidal zone Microcoleus-variations
Niveau level related to MHW	-130 cm	-10 cm	+30 cm	+30 cm
Reference to Fig.	37A	37B	37D	37F
MARINE FAUNA:				
Pygospio elegans	66	5.506	19.888	159
Corophium arenarium	4	452	1.908	0
Nereis diversicolor	43	232	74	1.302
Lumbricillus lineatus	1	603	1.986	19.462
Hydrobia ulvae	222	351	949	1.685
Bathyporeia sp.	60	4	35	
Capitella capitata	132	866		
Macoma baltica	56	64		
Heteromastus filiformis	88	94		
Eteone longa	20	18		
Scoloplos armiger	420			
Cerastoderma edule	2.057			
Arenicola marina	44			
Nephthys longosetosa	9			
Gattyana cirrosa	40			
Urothöe sp.	396			
Anaitides mucosa	28			
Nephthys hombergi	4			
Scolelepis bonnieri	16			
Crangon crangon	21			
TERRESTRIAL FAUNA:				
Bledius subniger			28	
Dolichopodidae larv.				348
Scatella subguttata				227
Heterocerus flexuosus				147
Bledius spectabilis				98
Species:	20	10	7	8
Individuals m^{-2}:	3.727	8.190	24.868	23.428
Shannon's diversityH':	2.45	1.75	1.06	1.01
Pilou's Evenness J:	0.57	0.53	0.38	0.34
Dominance (1-J):	0.43	0.47	0.62	0.66

A

B
7/270

C
MELLUM
4/105

D
7/281

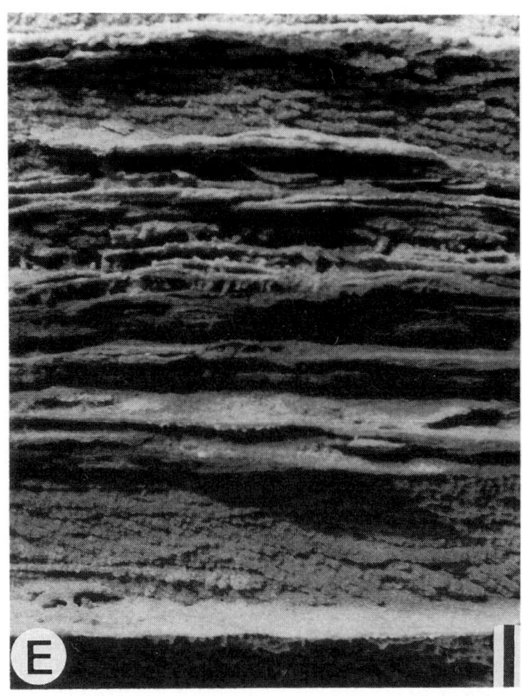

Fig. 37. Series of sediment cores showing from A - D the change in bioturbation structures vs increasing elevation and from D - F vs increasing microbial productivity. The corresponding fauna, studied with the reliefs at same levels and time, is listed in Table 10. Photographs by H.-E. REINECK. Scale is for all cores 2 cm.

A) Lower intertidal zone. Preexisting bedding is not visible, due to high grade bioturbation. Note life horizon of the cockle *Cerastoderma edule* beyond the surface.

B) Intertidal-supratidal zone boundary. Bioturbation is still high. Note, however, increasing visibility of tubes and burrows.

C) Same level as B. One single L-shaped trace, made by a juvenile lugworm, within evenly laminated sand. Initial stage of mat formation (see projecting top layer). The sample has been cored in the landpriel shallows.

D) Lower supratidal zone about 30 cm above MHW (*Oscillatoria* variation). High numbers of dwelling traces from intertidal animals pierce through sharply projecting buried mats.

E) Same level as D (transition towards the *Microcoleus* variation): Low numbers of dwelling structures reflect the substrate change, caused by increasing productivity of mats and decreasing sedimentation rates. In between the buried mats, physical structures (laminated sand, cross-bedding) are visible. Dark color indicates reduced sediments.

F) Same level as D (*Microcoleus* variation): The sediment is devoid of dwelling traces while roots of halophytes occur in great abundance.

The cockle *Cerastoderma edule* is a suspension feeder living just beneath the sediment surface. The maximum shell length is 5 cm (REISE, 1985). In response to sedimentation or erosion, the mobile population moves upwards or downwards to find its most appropriate settling depth, which is determined by the length of siphones (5 to 10 mm; see also the life horizon of *C. edule* in Fig. 37A). The high degree of activity of the population leads to the deformation of any newly sedimentated stratum. The polychaete *Scoloplos armiger* is a mobile burrowing bacterivore (REISE, 1985) which migrates through the sediment and leaves behind mottled bioturbation structures, clearly visible in Fig. 37A. Finally, the burrowing polychaete Arenicola marina is known as a very important bioturbator of sandy sediments. The number of 44 specimen per m^2 (Table 10) already represents a mean population density because of the large size of the lugworms (up to 30 cm). Annual sediment reworking amounts to a layer of up to 33 cm (CADEE, 1976).

As a consequence, the presence of this faunal assemblage which is dominated by the above-mentioned deformatively burrowing organisms leads to strongly bioturbated sediments without any visible preexisting sedimentary structure visible (Fig. 37A).

b) Upper intertidal belt

This belt shows already the above-mentioned relief unconformities merging into the sub-environments of mat formation which we have termed the *Oscillatoria* variation and the *Microcoleus* variation. The core in Fig. 37B indicates the base of the *Oscillatoria* variation. Increasing abundance of figuratively burrowing species (*Pygospio elegans*, *Corophium arenarium*; Table 10) gradually leads to the visibility of lebensspuren. The evenly laminated sand and the L-shaped trace of a juvenile lugworm in Fig. 37C indicate one of the landpriel shallows which cross the intertidal-supratidal zone boundary at the base of the *Microcoleus* variation. Juvenile lugworms benefit here from shallowest submersal conditions. Channel filling, however, followed by the development of a microbial mat (see top of Fig. 37C), may have had lethal consequences for the lugworm. WUNDERLICH's explanation of similar traces is, however that animals, removed from their natural habitat by accident (storm event?), have tried to escape from the sediment surface by burying and have produced these lethal escape traces (WUNDERLICH, 1984).

c) Lower supratidal zone (*Oscillatoria* variation)

It is notable that figuratively burrowing species which play a subordinate role in the lower intertidal base become more and more abundant with increasing height above MHW (Table 10). The dominant type *Pygospio elegans* (tube-forming) is associated with the burrowing amphipode *Corophium arenarium*, the mud snail *Hydrobia ulvae*, the burrowing polychaete *Nereis diversicolor* and the oligochaete *Lumbricillus lineatus*.

There is also a notable quantitative change: (1) the reduction in the number of species (7 vs. 20), (2) the increase in individual densities (25,000 vs. 4,000; mean values). Both changes go hand in hand with (3) a reduction of diversity (H'), (4) a reduction of eveness (J) and thus (5) the increase of percent dominance (1 - J). In trace-making species, the percentage dominance is:

Pygospio elegans	about 80 %
Corophium arenarium	about 8 %
Hydrobia ulvae	about 4 %
Nereis diversicolor (mainly juvenile):	about 0.3 %

The remaining 7 to 8 % belong to *Lumbricillus lineatus*, a 5-mm-long oligochaete. Traces of this species are not recognizable within sandy sediments. This might be different in silt and clay deposits where dwelling structures of oligochaetes are often detectable.

Hydrobia ulvae is the only one which produces deformative structures during low-tide resting (Fig. 36C). However, this species constitutes only 4 % of the total faunal assemblage, while animals which form agglutinated tubes and burrows reach a percentage dominance of about 90 %.

As a consequence, the occurrence of this faunal assemblage, which is highly dominated by figuratively burrowing organisms, leads to bioturbation patterns quite different from the lower intertidal zone (Fig. 37D). Suspension feeders (*Cerastoderma edule*) and animals with a large oxygen demand (*Arenicola marina*) are excluded. The high-lying area is appropriate for semi-sessile marine endobionts living in slime-agglutinated tubes and burrows. Physically and microbially derived laminations are visible, even though the population density of the invertebrates present is higher than in the intertidal area (Table 10). The various buried cyanobacterial mats visible in Fig. 37D actually represent time-lags where sedimentation did not proceed. This is the case during the summer period where flooding occurs less frequently. During the time span without sedimentation, the endobiontic animals benefit from the developing surface mat which supplies them with food and protects the underlying sediments from desiccation. With sedimentation going on again (wind is often the agent of deposition), the animals migrate upwards through each newly deposited layer and reinforce their prolongated dwellings by secreting slime. Former top mats, now buried, become perforated by the upwards movement of the animals. Nevertheless, the preexisting microbial laminae remain visible in the sediments (Fig. 37D).

3. 7. 2. Effects of increasing microbial productivity

The sediments of the *Oscillatoria* variation are characterized by a high population density of figuratively burrowing intertidal species. We calculated a mean distance of 6 mm between the dwelling structures (GERDES & KRUMBEIN, 1986). By contrast, the sediments of the *Microcoleus* variation contain lebensspuren of lower density (a maximum distance of 24 mm was calculated between burrows, mainly produced by juveniles of the intertidal polychaete *Nereis diversicolor*). Since both sub-environments lie at the same elevation level, there is no reason to imply the influence of decreasing flood recharge. Control of the abundance of burrowing and tube-building amphipods and small polychaetes of the *Oscillatoria* variation by the *Microcoleus chthonoplastes* mats and their degradational products is discussed in the following:

Five marine intertidal species occur in both sub-environments of mat formation but show considerable changes in dominance and abundance (Table 10). This can be correlated to substrate differences. As mentioned before, the *Microcoleus* variation (protected shallows) can be distinguished from the *Oscillatoria* variation (open less protected slope) by (1) lower rates of sedimentation; (2) higher standing crops; (3) higher concentrations of reduced metabolic compounds (NH_4^+, S^{2-}); (4) strongly negative Eh-values and lower pH values. Besides slow sedimentation, all other parameters are microbially induced, while grain-size distribution in both variations is nearly the same. Slow sedimentation contributes on the other hand indirectly to the substrate change since layers rich in organic carbon are closer together.

The substrate change is accompanied by a change in dominance between trace-making species and those which leave no traces behind. In trace-making species, the percentage abundance is:

Pygospio elegans:	about	0.7 %
Corophium arenarium:	+/-	0 %
Nereis diversicolor (mainly juvenile):	about	4 %
Burrowing beetles	about	3 %

Thus the mean percentage abundance of trace-makers in the *Microcoleus* variation amounts to 8 % while in the *Oscillatoria* variation it reaches 90 %. The small oligochaete *Lumbricillus lineatus* constitutes 85 % of the total faunal assemblage. The remaining 7 % belong to *Hydrobia ulvae* (Fig. 36C) and diptera. The latter also leave no internal traces.

It is notable that the total abundance of individuals x m^{-2} is nearly the same in both variations (Table 10). The change in species domination between the sub-environments of mat formation accounts for the biofacies change detectable in horizontal direction (see relief casts D, E and F in Fig. 37). While the sediments of the *Oscillatoria* variation are marked by a high density of trace-making species, the *Microcoleus* variation is comparatively void of faunal traces, as a result of about 85 % dominance of the oligochaete *Lumbricillus lineatus* which does not leave any trace in sandy sediments. *Lumbricillus lineatus* prefers the intermediate position between the top layer of *Microcoleus* mats and the highly anoxic subsurficial siliciclastic sediments where it feeds on bacterial and diatom coatings of sand grains (GIERE, 1975).

The examples provided here show that bioturbation grades and patterns are effectively controlled by changes in dominance and abundance of animals with different life habits.

3. 7. 3. Promoting and limiting distributional factors

Two questions should be discussed with respect to the observed distribution: (1) What are the factors promoting marine burrowing invertebrates to persist within an area of irregular flooding? (2) What are the factors preventing these organisms from burrowing in the thickened biolaminated deposits?

The first question may be addressed to the "opportunism" phenomenon. The most abundant trace-making taxon is the small polychaete *Pygospio elegans*. The mode of reproduction identifies this species as a typical opportunist (GALLAGHER et al., 1983; GERDES & KRUMBEIN, 1985). The larval development usually includes a pelagic stage which is covered by the tides. The life-cycle of this species can, however, be exclusively benthic. In this case the products of reproduction are laid directly into the tube along with nutrient "eggs". The developing larvae spread out and colonize the area in close proximity to the parents without entering a pelagic stage (SMIDT, 1951). This species is also capable of asexual reproduction, and its increase on defaunated flats has been observed to be almost exclusively due to the asexual reproduction mode (RASMUSSEN, 1953, 1973; HOBSON & GREEN, 1968; GALLAGHER et al., 1983). This seems to be an ideal tactic to persist on the high tide flats of Mellum, and in fact, we observed asexual stages within the samples obtained from Mellum.

Characteristic of opportunists such as *Pygospio elegans* and possibly also *Corophium arenarium* (DAUER & SIMON, 1976) is thus a variety of methods, high levels of reproduction and short life spans of the individual. It is these groups which usually re-establish themselves first after catastrophic declines in population caused, for example, by increased deposition, erosion or freezing in winter. After the substrate conditions have been re-established they are often superseded by those populations which were temporarly destroyed. Such mechanisms of supersession are impossible at the supratidal flats of the Farbstreifen-Sandwatt. Consequently populations of these species first able to colonize this area remain constant and can - almost without competition - become denser than intertidal populations. The series of events which led to the increase in numbers of the snail *Pirenella conica* in the Gavish Sabkha and of the opportunistic polychaetes and amphipodes in the Farbstreifen-Sandwatt seems to be related: "Environment produces facies" (TEICHERT, 1958).

The second question should be addressed to chemical conditions prevailing in the *Microcoleus* variation. The field data elaborated in this study give some indications that there is a coupling effect of low-rate sedimentation and high standing crops realized in this variation. This coupling effect results in a pile of reduced organic carbon and reduced intermediate products due to bacterial break-down within an otherwise clean, mainly wind-borne depositional environment of quartz-sand. The *Microleus* variation thus may represent a sub-environment where chemical conditions associated with the mat productivity minimize the density of trace-makers. Studies in the laboratory in order to explain and predict trace-scarcity with an increase of degradable organic matter from microbial mats (GERDES & KRUMBEIN, 1986) have revealed the following results:

1. Ammonia as a limiting factor: Both, the burrowing amphipode *Corophium arenarium* and the tube-forming polychaete *Pygospio elegans* tolerate ammonium contents of relatively high levels. Values of ammonia concentrations measured within sediments of the *Microcoleus* variation lie generally below the tolerance limits of both species. However, GRAY (1984) suggests a confidence range of about a tenth of the actually found tolerance limit, if only one single factor such as ammonia is considered, since sediments rich in organic matter comprise a multitude of complexly interacting processes and metabolic substances.

2. Oxygen deficiency as a limiting factor: Animals of the species *Corophium arenarium* die very soon when placed in sediments of low oxygen (< 2 mg/l). *Pygospio elegans*, on the other hand, tolerates oxygen depletion in pore waters. The latter species is also known to live in anoxic intertidal sediments.

The sensity to oxygen deficiency explains why *Corophium arenarium* is not able to live in sediments of the *Microcoleus* variation (compare Fig. 34C). These sediments restrict, however, also the tube-forming polychaete *Pygospio elegans*, while at equate level (same inundation frequency) high population densities occur. Hence, a multifactorial effect (enrichment of reduced compounds, stagnant interstitial water during periods of subaerial exposure and high energy costs of feeding on the tough surface mats) may be responsible for the limitation of this important tracemaker within thickening microbial mats.

3. 8. Intertidal-supratidal sequence

The decreasing influence of tides considered (1) from the hydrodynamic and (2) the ecologic point of view is congruous with the "sum of modifications" (GRESSLY, 1838) in internal and surface structures vs elevation. The following sections summarize the patterns of change in internal structures (Fig. 38) and surface structures (Fig. 39) along the intertidal-supratidal sequence on Mellum Island.

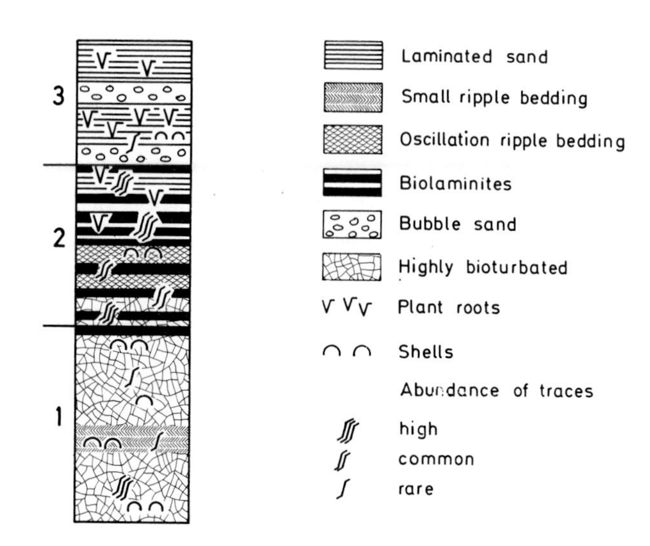

Fig. 38. Idealized vertical succession showing the prograding intertidal-supratidal sequence and the position of microbial mats at the intertidal-supratidal zone boundary. Section 1: intertidal zone, section 2: lower supratidal zone, section 3: upper supratidal zone.

3. 8. 1. Change of sedimentary internal structures

High grade bioturbation controls most parts of the lower and upper intertidal zone. With the exception of some current ripple structures, preexisting bedding is not visible. The extensive burrowing reflects a diverse and mobile infauna which dominates over current and wave stratification.

At a level which corresponds to the intertidal-supratidal zone boundary, a change in appearance of physical and biogenic structures becomes obvious (Fig. 38):

(1) The physical structures include parallel-laminated sand, small scale cross-bedding, resulting mostly from wave ripples, and bubble sand (secondary physical structure; REINECK & SINGH, 1980), all characteristic of sandy tidal flats exposed to wave action. The parallel-laminated sand which is the dominant physical structure may, however, be produced mainly by wind, though REINECK (1963) attributed it also to swash and backwash wave action.

(2) Buried mats appear at a level which corresponds to the MHW-level. Each mat represents a former surface which was covered with sand loads and was subsequently bound by microbial activity. The more the sequence grades upward into the supratidal environment, the more mat generations appear. The spaces between the individual mats decrease with increasing height above MHW. At a level which corresponds with the boundary between the lower and upper supratidal zone (i. e. between 40 and 50 cm above MHW), the mat horizons gradually disappear and physical structures (laminated sand and bubble sand) dominate (Fig. 38).

(3) The appearance of individual traces (tubes and burrows) reflects a complex change of infaunal assemblages. Among the tubes and burrows are root shafts of *Salicornia* sp., which also indicate supratidal conditions (Fig. 38). Species which cause the extensive burrowing in the intertidal sediments are excluded by irregular flooding.

3. 8. 2. Change of sedimentary surface forms (Fig. 39)

The intertidal zone is characterized by diverse ripple systems indicating the influence of tidal currents. Around the intertidal-supratidal zone boundary smooth patches appear which alternate with ripple systems. The smooth patches indicate microbial colonization going hand in hand with the immobilization of surface sediments. The intertidal-supratidal zone boundary is, however, a struggling zone where flooding frequencies are still sufficient to interfere with microbial activity (REINECK, 1979). Looking back to the non-colonized lower parts of the intertidal zone, we see, however, clear modifications of the surface relief: The ripple systems are confined to pockets which indicate the partial erosion of microbially immobilized surfaces and subsequent rippling of the bare sand within the pockets. Wave ripples are dominant within the erosional pockets. With further in-

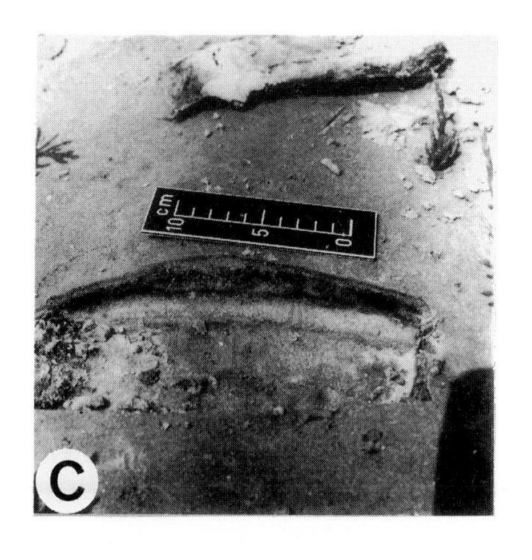

AREA
AROUND
MHWS

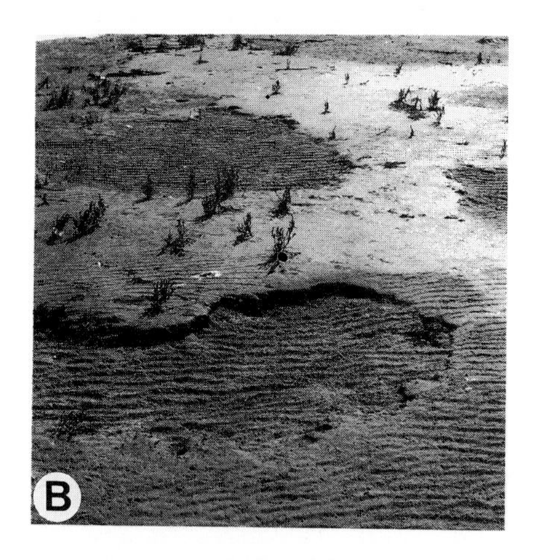

INTERTIDAL-
SUPRATIDAL ZONE
BOUNDARY

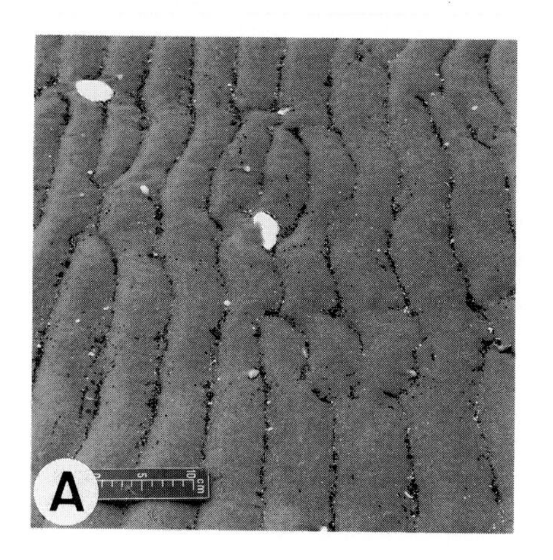

INTERTIDAL
BASE

Fig. 39. Change of surface structures vs elevation. Photographs by H.-E. REINECK.
A) Intertidal base, characterized by extended ripple systems.
B) At the intertidal-supratidal zone boundary: Ripples are confined to flutes where microbially immobilized surfaces became eroded. At the bottom: Freshly deposited sand layer on top of a mat shows more extended ripple systems. (Modified after REINECK, 1979).
C) At the level of MHWS: Smooth surfaces predominate. In the vertical section visible are several microbial mat generations which lie beyond a thin air-borne layer of sand.

increase of lower supratidal slope the smooth patches increase in number and extension and gradually replace the erosional marks (Fig. 39C).

Air-borne sand deposited on surface mats can be effected by wave action or currents before it is recolonized and secured by the buried microbiota or from outside. In this case, the unsecured sand is displaced and ripples can form (see Fig. 39B). If this is followed by a period of calm weather without further movement of the sediment, then a new mat forms on the rippled surface, following the morphology of the ripples. When deposition reoccurs, the mat is buried, retaining this wavy characteristic (see Fig. 33C). Although the sideways expansion of the ripples within erosional pockets is restricted to a few decimeters by the size of the pockets, the rippled and newly overgrown carpets of air-borne sand extend over a larger area.

Interpretation. The critical speed of water flow at which transportation of fine-grained sand begins lies between 0.2 and 0.4 m/s. This critical point increases in biologically secured sand on average by $V_{crit.} = 0.98$ m/s and a maximum of $V_{crit.} = 1.56$ m/s (MANZENRIEDER, 1984). Biologically secured sand is only eroded during bad weather with a combination of high water levels and movement of the sea and especially where surface irregularities such as broken shells provide a starting point for erosion.

Example from the fossil record. In the cretaceous Dakota Group at Alameda Avenue west of Denver/Colorado a fossilized bedding plane (MACKENZIE, 1968, 1972) with eroded structures has been revealed which closely resembles the surface of the "Farbstreifen-Sandwatt" (REINECK, 1979). This outcrop at first evokes the impression that the ribbed surface of the flute may belong to a lower-lying bedding plane and have been revealed by weathering. The recent example of the Farbstreifen-Sandwatt, however, shows that partial erosion of sandflats already secured by microorganisms may produce these structures (Fig. 39B). In both recent and fossil examples the ripple crests in the erosion pockets blend into the smooth surface edges. This characteristic indicates that the flute and the ripples within it were formed after the flat surfaces (REINECK, 1979).

3. 9. Subaerial rise of biolaminated quartz-sand
(experimental approach)

The depositional record shows that sand layers are sandwiched between layers of organic carbon-rich horizons, due to microbial mats which proliferate during intervals of non-deposition. Bed-by-bed recognition within a texturally and structurally uniform lithologic sequence thus becomes possible. In a similar rock record, however, it may not be easy to infer completely subaerial conditions involved in the raise of the biolaminated quartz-sand. Thus, the question may arise: Is it possible without any help of tides, or is at least a periodical flooding necessary in order to re-inoculate freshly sedimentated layers with microbes held in suspension?

We studied this question in the lab (Fig. 40): (1) A core was excavated in the field (*Microcoleus* variation) with a plexiglas-cylinder 40 cm in diameter and 40 cm high. The filled cylinder was set into an aquarium tank and the space between the cylinder and the tank walls was filled with fine-grained sand. The sand was filled up to about one half of the height of the cylinder. The sediment inside the cylinder was not flooded, but capillary action was aided by pouring seawater on the sand outside the cylinder. (2) A 5-mm layer of fine-grained sand (100 µm grain size) was scattered over the mat surface within the cylinder with a sieve. This simulated low-rate wind-laid deposition. (3) Low-rate sedimentation was repeated every four days, while in the meantime the development of new mat was visually observed and measured by pigment concentrations. (4) The experiment was finished after twelve instances of oversedimentation, and a relief cast was prepared (Fig. 40).

The test has shown that a multitude of microbial mat generations can develop from one and the same preexisting mat without any inoculation by microbes held in suspension by the tide waters. Apparently, wind-

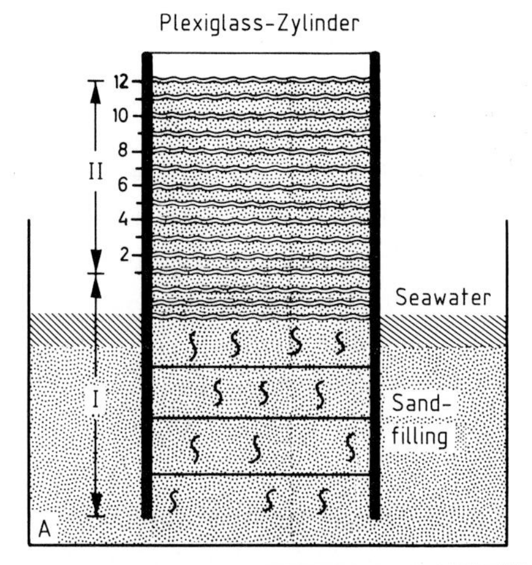

Fig. 40. Subaerial rise of biolaminated quartz-sand (experiment).
A) Treatment: Undisturbed sediment with a mat on top was cored in the field (section I). Twelve instances of oversedimentation (section II), carried out in the lab, simulated low-rate wind-laid deposition without flooding.
B) Relief cast preparation showing the twelve generations of microbial mats that originated from one and the same preexisting mat. The test has shown that biolaminated sequences can grow without flooding and inoculation by microbes held in suspension by tide waters. Photograph by H.-E. REINECK.

borne low-rate sedimentation acts as a "stimulant" which aids the subaerial rise of biolaminated deposits in a rarely flooded high tide flat provided that moisture is sufficient due to capillary action.

The structures evolving from these processes are continuously horizontal and regularly interlaminated. No diversity of microbially produced sedimentary structures occurs as in the Gavish Sabkha, where microbial communities rich in different biotypes are facilitated by persisting shallow-water environments on the one hand and long-term periods of nondeposition on the other hand.

3. 10. Summary and conclusions

Cyanobacteria account for biogenic sediment accretion in the "Farbstreifen-Sandwatt" (versicolored sandy tidal flat) of Mellum Island. The biolaminated deposits are similar to facies type 1 of the Gavish Sabkha (siliciclastic biolaminites). Differences exist in the lithology: Sediments upon or through which the mats on Mellum Island grow are made up of clean sand. The grains originate predominantly from reworked glacial sediments and are rounded to well rounded. By contrast, the strong angularity of siliciclastic grains in the Gavish Sabkha clearly shows their status as primary weathering products.

Dominant are filamentous cyanobacteria representing the same genera and in the case of *Microcoleus chthonoplastes* also the same species as in the Gavish Sabkha and the Solar Lake. Coccoid unicells, however, contributing in the hypersaline environments with excessive gel production to translucent, vertically extended laminae are less important. This may be due to the general lack of shallow-water conditions.

Due to the dominance of filamentous cyanobacteria, the biolaminites are continuously stratiform. Episodic low-rate sedimentation stimulates the mats to grow through the freshly deposited sediment layer. Wind is the major agent of sediment transport. Modifications of the microbial community structure, standing crops, redox potentials and pH are highly correlative to different degrees of protection realized by relief unconformities in the lower supratidal zone.

There is no indication that burrowing and grazing by intertidal and terrestrial animals are detrimental to the growing mat systems. According to the marine fauna, two distributional barriers exist: (1) physical and (2) biogeochemical factors. Intertidal deformative burrowers such as cockles and lugworms are controlled by the physical barrier of decreasing regularity of flooding. In spite of this fact, the mats are pierced through by numerous dwelling traces. These stem from small polychaetes and amphipode crustaceans which are able to spread over the intertidal-supratidal boundary and settle up to the MHWS-level. Biogeochemical barriers are: Oxygen depletion within the sediments, high ammonia and sulfide contents, which generate through bacterial breakdown of organic matter. Within the highly productive mats of *Microcoleus chthonoplastes*, dwelling traces of marine polychaetes and amphipode crustaceans disappear due to these conditions. The name of the mat-forming species, *Microcoleus chthonoplastes*, indicates its capacity to form "soils" (Greek chthonos). While lithology is not altered, the presence of *Microcoleus* mats leads to a habitat change which excludes trace-making "arenophile" invertebrate species and favors "chthonophile" species. The latter include high densities of individuals which do not leave traces behind.

In summary, the degree of self-organization of biolaminated deposits in the temperate climate zone is lower than within the hypersaline, protected shallow water basins of the Gavish Sabkha and the Solar Lake. The biolaminated deposits grow with the aid of quartz-sand sedimentation. In correlation with the capacity of the microbial mats to reestablish themselves on freshly sedimentated surfaces, the sedimentation rate per time unit is highly responsible for the amount of organic matter stored within the sediment column and indirectly it is also responsible for the density of traces made by marine burrowing metazoans.

4. WHAT THE ENVIRONMENTS HAVE IN COMMON - FINAL REMARKS

This part on modern stromatolite environments may be finished by considering three attributes they have in common: the presence of cyanobacteria, the trace record relating to salt beetles and the position on the margin of a sea.

4.1. Cyanobacteria as pioneer organisms

Cyanobacteria are well adapted to peritidal conditions. They are capable of surviving extremely high salt concentrations, high temperature, strong insolation, extremely low water potentials as well as drastic short-term changes in these conditions. For these reasons cyanobacteria have apparently survived for 3.5 milliards of years and continued to thrive through Earth history.

A remarkable attribute of cyanobacteria occupying sandy high tide flats such as those of Mellum Island is their capacity to fix molecular nitrogen. Fertilization of the clean (mainly wind-borne) and thus nutrient-poor quartz-sand gradually takes place through physiologic pathways of nitrogen reduction, accumulation of organically bound nitrogen within the densely populated mats and its consequent release by chemotrophic bacteria. This may signify a mode of facilitation for the subsequent development of macroalgae and macrophytes. The nitrogen-fixing enzyme system is, however, very sensitive to O_2 and, for several species, is protected in separate cells, called heterocysts. STAL et al. (1984 a, b, c) recognized, however, that cyanobacteria inoculating the high tide flats of Mellum Island do not possess heterocysts, but are nevertheless active in nitrogen fixation. One of the major mat-forming species, *Oscillatoria limosa*, is of this type.

The similarity to widely accepted theories about the evolution of the oxygenic photosynthesis is striking. Early in Proterozoic time, microorganisms which may have been cyanobacteria-like may have liberated themselves from anoxygenic photosynthesis with H_2S as the electron donor and developed systems which enabled them to use the universal H_2O as an electron donor, and O_2 was formed as a by-product. This change represents probably the most drastic event in the history of earth, since it acted as a mode of facilitation for fungi, plants, animals and man. The nitrogen-fixing bacteria, however, had to pay for the emancipation from anoxygenic photosynthesis by having to protect the O_2-sensitive enzyme system under changed conditions (STAL & KRUMBEIN, 1981; STAL et al., 1984 a, b, c).

4.2. Saltbeetles: "Purpose" of dwelling burrows

The information revealed by traces about their builder is of paleontological interest. In rock a trace is often all that remains of the sediments' inhabitants. SEILACHER (1951) showed that with a specimen of tube-building marine polychaete we can learn much from the construction of its tube about its "ecological purpose" - the function of the construction for its inhabitant in a particular environment - and perhaps thereby find indicators for the environmental conditions prevailing at the time of deposition.

Contact with seawater is essential for the survival of marine gill-bearing organisms. Such contact can endanger the life of terrestrial air-breathing organisms which contribute in all three environments studied to the lebensspuren spectrum. The "purpose of dwelling burrows" for terrestrial organisms living in an aperiodically flooded depositional environment lies in the fact that (1) long-term contact with seawater is to be avoided and (2) vital needs can be satisfied without external contact for a certain period of time.

The morphology of dwelling burrows of the salt beetles (*Bledius angustus* and *B. capra* in the Gavish Sabkha, *B. subniger* and *B. spectabilis* on Mellum Island) shows that these constructions serve in the protection, sustenance and way of life of their builders in a habitat which irregularly floods and dries out. They are obviously (1) biologically dependent for their sustenance on the presence of photosynthetic microorganisms; (2) they could to a large extent avoid the unfavorable consequences of living on the interface between land and see through their highly specialized methods of dwelling; (3) with these characteristics they are predestined to live in a depositional environment where microbial mats predominate. Examination and interpretation of their lebensspuren in a paleoecological sense is currently theoretical, as directly comparable fossil traces from stromatolitic deposits are still lacking today.

4.3. Peritidal settings

All the environments studied are situated on the margin of a sea. Within a hypothetical stratigraphic succession they would take almost the same position, overlying a set of intertidal sediments and merging into upper supratidal sequences. Gavish Sabkha and Mellum Island describe, however, two different vertical profile models.

4. 3. 1. The "sabkha cycle"

Coastlines of subtropical, arid regions prograding seawards follow the dynamic sequence of the "sabkha cycle". The SE incline of the Gavish Sabkha center also shows this prograding tendency. This is also characteristic of the Persian-Arabian Gulf (SHEARMAN, 1978; KENDALL, 1979). There in the sequence from the land to the sea can be found stretches of (1) widespread supratidal areas of halite, anhydrite and gypsum deposits, (2) transitional zones from supratidal to intertidal zone of carbonate sediments, mainly dolomite, and (3) subtidal half-closed back-barrier lagoons. Multi-layered microbial mats in association with ooids and oncoids as well as the nodular Pleurocapsalean aggregates are confined to the carbonate sediments of the upper intertidal zone. On higher levels they are replaced by sulfate deposits.

The dynamic sequence of the "sabkha cycle" which ends with terrigenous sediments can be used as a recent model to explain "paleosabkha cycles". Cyclically reoccurring stratigraphic sequences can be observed in the Narssarsuk formation in Northwest Greenland (Precambrian). Each cycle begins with limestone, which is succeeded vertically upwards by organically rich dolomite. This facies is interfingered with gypsum higher up and finally is covered by fine-grained red sandstone (STROTHER et al., 1983; KNOLL, 1985a). The banded dolomite sequence contains microfossils of coccoids and thread-like organisms as well as *Eoentophysalis* aggregates (KNOLL, 1985a). The morphological similarity of this microfacies with the recent Pleurocapsalean aggregates of the Gavish Sabkha has already been discussed (Facies type 2).

Similar paleosabkha cycles can be found in the most varied geological formations. In all cases stromatolites are found in association with carbonates and are replaced by sulfate deposits in overlying strata (FRIEDMAN & SANDERS, 1978; PURSER, 1980; SHINN, 1983).

The re-establishment of stromatolitic carbonates above sabkha deposits can be seen in vertical profiles of the Solar Lake sediments, caused by a tectonic event. In other cases transgressions may have been responsible (see in Part III the Permian Zechstein sequence).

4. 3. 2. Temperate humid coastlines

In the temperate humid coastal zone, horizons enriched in iron oxides and iron sulfides, both occurring sometimes within one and the same sediment column, may be viewed as the "equivalent" for the microbial mat-mediated carbonate mineralogy in the arid coastal zone. If sedimentation of siliciclastics is frequent as on Mellum Island, the vertical set is confined by a regular interlayering of organically rich (and iron sulfide-rich) and organically poor horizons. Varied studies along the North Sea Coast have shown that this kind of growth-bedding within a texturally uniform sand is widely distributed (POTTS et al., 1978; MEYER & MICHAELIS, 1980; COLIJN & KOEMAN, 1975; HAUSER & MICHAELIS, 1975; REINECK & GERDES, 1984; SCHULZ, 1936; HOFFMANN, 1942; REINKE, 1903). The zones where these structures develop are (1) level uniform, (2) low-energy environments, (3) made up of fine-grained quartz sand with very low contents of silt and clay, (4) initially low in nutrient concentrations. Their occurrence, however, ranges from backsides of barrier islands to stationary sand banks, from beaches of barrier islands to those of mainland coasts. Consequently, biolaminites overlie tidal flat and beach sequences, they merge with dune cross-bedding or rhizospheres of salt marshes, and they bear a different faunal composition: mixed marine-terrestrial on barrier islands and mainland coasts, merely marine on sand banks without connections to terrestrial habitats. Wherever these structures develop, their characteristic features are repeated: Smooth surfaces alternate with small erosional pockets, laminated sand, sometimes also small-scale ripple bedding with dark laminae rich in organic matter. The vertical sequence does not show any changes of grain fabrics, which could imply that the organic matter resulted from slow sinking deposition from suspension (e. g. tidal bedding).

The characteristic appearance of biolaminites in the lower supratidal zone of various geomorphic units led us to characterize these deposits as the biogenic laminite subfacies (GERDES et al., 1985c).

PART III

SPANNING THE GAP BETWEEN MICROBIOLOGY AND GEOLOGY

"Up until now I have concentrated my attention on those bodies which were discovered in places which give us cause to doubt that they were their places of origin and thereby I have shown how from the perceptible we can draw unperceptible conclusions." (N. STENO, 1667)

1. INTRODUCTION

In sedimentary rocks of the Proterozoic era microfossils were found which are considered to be the signs of the earliest forms of life on earth. For many fossil discoveries a formational environment similar to present-day hypersaline coastal environments is assumed. This assumption is based on the high level of morphogenetic similarity between the fossil and recent stromatolites and their associated forms. Recent systems are then taken as models for the genesis of facies (KNOLL, 1985a), and the statement "the present is the key to the past" (GEIKIE, 1905) is also relevant in the field of paleomicrobiology which spans the gap between geology and microbiology. The converse of this statement is, however, also meaningful. In many cases it was geology which first opened our eyes to the similarity of recent forms to those ancient ones and consequently encouraged research into them.

Spanning the gap between microbiology and geology means (1) matching of microfossils in sedimentary rocks with experience from modern microbial morphotypes since cellular structures (and most probably also metabolic pathways) of procaryotes show a remarkable constancy throughout geological time (CLOUD, 1976); (2) matching of microstructures (e. g. of microcolumnar shape) with experience from processes which bring about these structures in modern microbial ecosystems; (3) matching of mineral precipitates with experience from modern microbially induced mineralization mechanisms; (4) matching of associated sedimentary structures with experience from modern depositional environments which favor the establishment of microbial communities.

The direct comparison of ancient and modern stromatolites necessitates, however, some precautions: Several structures resembling stromatolites are abiogenic (e. g. caliche). Postdepositional contamination of weathered outcrops (BUICK et al., 1981) causes problems since organisms of microscopic scale are able to penetrate the smallest fissures and often are lithified by secondary intrusions (for further discussions see DUNLOP et al., 1978; CLOUD & MORRISON, 1979; SCHOPF & WALTER, 1980; BUICK, 1984).

The combination of the above-listed single specific observations may allow for the reasoning that the observed structures, mineral deposits and particles (e. g. ooids and oncoids) are (a) biogenic, (b) formed in situ, (c) indicate specific formational processes in interaction with the external environment which may shed light on the type of depositional environment and probably also on paleoclimatic conditions.

In the following chapter, we present studies on microbial structures in rocks. One example is given of platy dolomite in the Permian Zechstein sequence (PZ3) of North Poland which might allow the inference of a depositional environment similar to the above mentioned present-day coastal sabkhas. The other examples given (Precambrian Gunflint iron formation, Canada and Lower Jurassic ironstone, Lorraine) may widen the scope inasmuch as the microfossils found imply that metabolic types others than cyanobacteria have been involved in the formation of stromatolites and ooids and in the concentration of minerals. The final chapter is a complementary summary of pathways involved in microbial sediment accretion.

2. METHODS

Examination of thin sections under the light microscope are often insufficient for the recognition of microfossils. As in other cases, the SEM made a decisive breakthrough. For example, KAZMIERCZAK & KRUMBEIN (1983) - after an initial combination of fracturing and etching with different levels of acidity and subsequent SEM analysis - could demonstrate that the stromatoporoid structures of Silurian rocks in the Wenlock Stufe, N. E. Gotland, was formed by coccoid microorganisms with multiple division similar to the recent structures in the Gavish Sabkha.

The methods described by the said authors were adopted for the following studies. Selective etching was preferred, i. e. producing a comparison in relief between two or more minerals species. We used a weak acid solution (1 - 5% HCl) to reduce reliefs of calcite in interlayered carbonate rocks and to maintain reliefs made up of dolomite and other minerals insoluble in HCl. Parallel thin sectioning of rock specimens and fracturing of selected chips for SEM was chosen because of the orientation of microstructures obtained by SEM. Thin sections were prepared and examined under dissecting and polarizing microscopes, selected chips were fractured, etched stepwise and subsequently coated with carbon and/or Pt for SEM studies (Steroscan-180, Cambridge Instrument).

The samples studied come from (1) Precambrian Gunflint iron formation, Ontario/Canada, (2) Permian Zechstein of North Poland (PZ3), (3) Lower Jurassic Minette ironstone of Lorraine. The basically sedimentary origin of the rocks has been proved by previous researchers.

3. DESCRIPTION AND INTERPRETATION OF FOSSIL MICROSTRUCTURES

3. 1. Precambrian Gunflint iron formation, Ontario

3. 1. 1. Provenance of rock samples and previous work

The Gunflint formation extends from the west of Gunflint Lake to Thunder Bay about 180 km to the east and continues towards the northern shore of Lake Superior just west of Schreiber, Ontario (BRODERICK, 1920; GOODWIN, 1956; GOVETT, 1966; GARRELS et al., 1973; MARKUN & RANDAZZO, 1980). The above-listed authors suggested environmental settings including the shallowest marine conditions within marginal basins, probably with restricted connections to the sea (TYLER & TWENHOFEL, 1952; JAMES, 1954). Sediments were deposited 1.9 - 2.0 Ma ago (KNOLL et al., 1978) and contain shales, chert-taconites, bedded chert carbonates overlying a basal black chert. Microfossils have been recognized in stromatolitic buildup (BARGHOORN & TYLER, 1965; AWRAMIK & BARGHOORN, 1977; KNOLL et al., 1978). The rock specimen studied here comes from the basal black chert.

Fig. 41. Microstructures and microfossils from the Precambrian Gunflint iron formation, Ontario.
A) Stromatolitic microcolumns form pockets, depressions and cavities containing ooids and oncoids of various shape and size. Scale is 2 mm.
B) Abundant filamentous microfossils preserved within dark stromatolitic laminae. Scale is 200 µm.
C) Filamentous microfossils within light laminae. Scale is 200 µm.
D) The nucleus of the faint concentric structure shows a meshwork of filaments (probably intraclast). Left: Oncoid. Scale is 200 µm.
E) Outer coating of an ooid showing filaments in concentric arrangement. Scale is 25 µm.
F) Modern microcolumnar build-ups (vertical section) from saltworks (Bretagne) are shown for comparison. Scale is 5 mm.

3. 1. 2. Microstructures

Description: The lower zone of the basal black chert of the Gunflint iron formation shows microcolumnar stromatolites, built up by alternating dark and light-colored laminae (Fig. 41A; AWRAMIK & SEMIKHATOV, 1979). Filamentous microfossils are abundant within the dark and less abundant within the light-colored laminae (Figs. 41B, C). The morphology of the microfossils is similar to cyanobacteria, although a close morphological similarity exists also between the Gunflint filaments and modern iron bacteria (CLOUD, 1965; KNOLL & AWRAMIK, 1983). Single and compound ooids of various shapes and sizes (250 µm to 4 mm in maximum diameter) occur within the light-colored laminae and in pockets, depressions or cavities between the microcolumns (Fig. 41A). They are associated with intraclasts probably mediated by former cell clusters or fragmentated microbial mats. Intraclasts as well as nuclei of ooids (Fig. 41D) show meshes of filamentous microfossils similar to the dark laminae. The matrix around the different particle species consists of microcrystalline siliceous and carbonaceous material. Hematitic and pyritic iron mineralizations also occur. The mineral coatings around the ooids are alternately dark (siliceous/ferruginous) and light (siliceous). Filamentous microfossils occur also in the outer coatings and show concentric arrangements (Fig. 41E). The ooids are more abundant towards the bottom of the small depressions which were produced by the microcolumnar arrangement of the stromatolites.

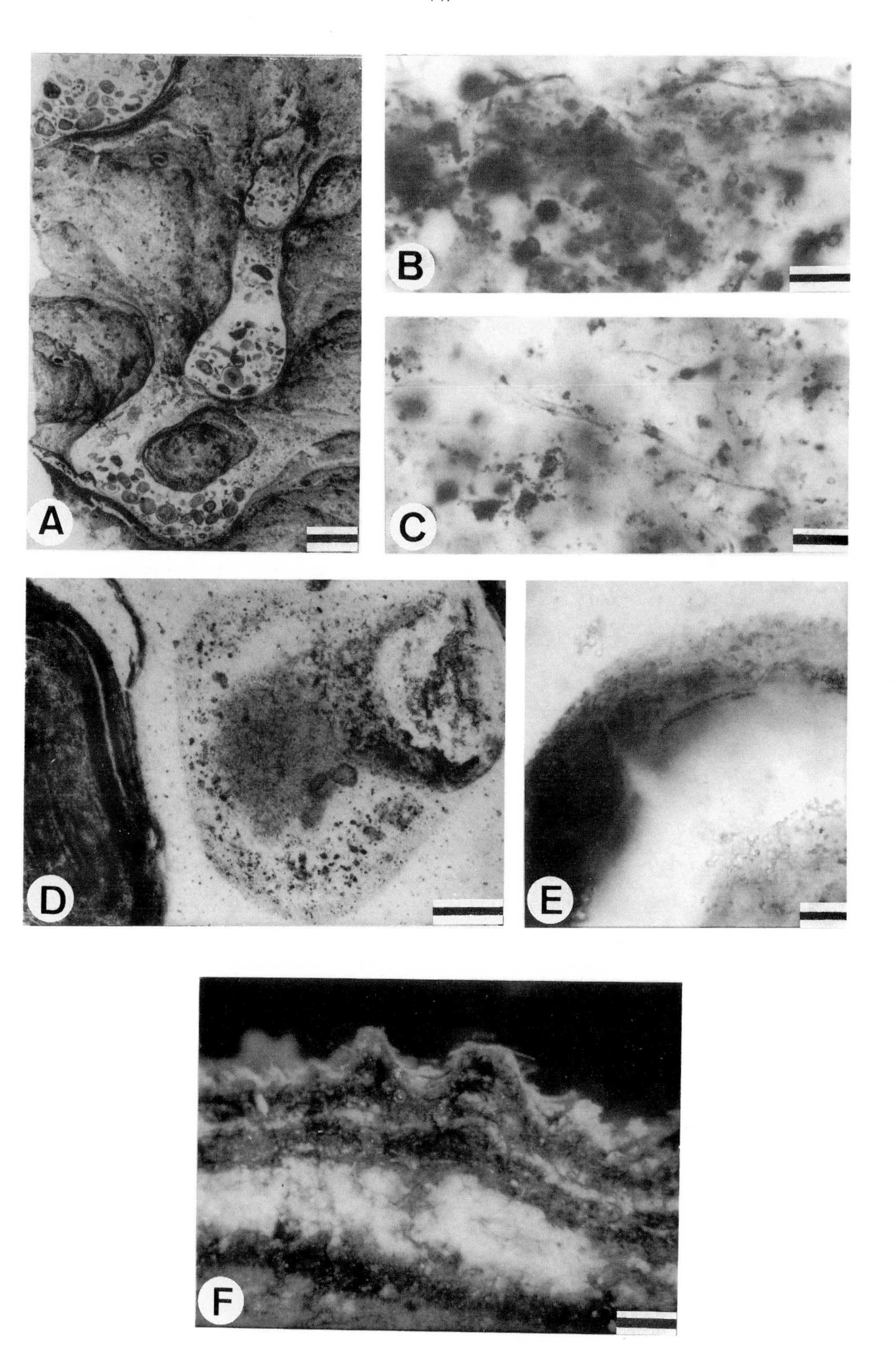
A
B
C
D
E
F

Interpretation: The development of microcolumns is known in various present-day microbial mats, referred to as pinnacles. Pinnacle mats are most abundant in low constantly submerged basins, such as in the lower part of the Solar Lake shelf. The pinnacles, emerging from the filamentous laminated surface mat, are often filled with oxygen and coated at the outer surface with slime. Their vertical extension ranges from 1 mm to about 5 mm. Recently we have observed these features in salt works where seawater supply is regulated to constantly cover the mats. A mild hypersalinity favors the production of slime, which covers the outer surface of the pinnacles and accumulates in interspaces (Fig. 41F). Mucilage partially diluted in seawater and floating aggregates of unicellular cyanobacteria occur also in the surface water and accumulate in between the pinnacles due to gravitational settling.

The pinnacle mat may be the most appropriate model for Gunflint microcolumnar buildup, the cavities and pockets in between and even the appearance of single and compound ooids, oncoids and intraclasts within the pockets and cavities. In the light of modern observations, the pockets and their fillings of masses of slime are ideal prerequisites for the localized precipitation of authigenic minerals around highly reactive decay centers, maintained by intraclasts and cell aggregates.

3. 2. Permian Zechstein Plattendolomite of North Poland (PZ3)

3. 2. 1. Provenance of rock samples and previous work

Rock specimens from the Leba Elevation, N. Poland were studied in collaboration with A. GASIEWICZ, who also collected the specimens. GASIEWICZ (1984) proposed a shallow water carbonate platform which underwent a very slow but progressive deepening where sedimentation rates were high enough to compensate for the rate of transgression.

3. 2. 2. Microstructures

Description: The Plattendolomite sequence of the Leba Elevation of N. Poland represents the peripheral part of the southern Permian basin. It is characterized by dolomites, limestones and anhydrite (PERYT et al., 1985). The lower and upper parts of the sequence bear regularly laminated structures (stromatolite-like or stromatoloid sensu BUICK et al., 1981). The middle zone, about 12 m thick, shows a relatively uniform character with deposits which bear poor or indistinct lamination. The following description deals with the fabrics and microfossil record of rock specimens of this middle section which we studied in cooperation with A. Gasiewicz (GASIEWICZ et al., in preparation). Material from the lower and upper sections is currently under analysis and will be presented later.

The middle zone of the Plattendolomite sequence displays a wavy, horizontally oriented lamination of alternating thin, commonly discontinuous dark laminae (about 1 mm thick) and more extended (up to 2 cm thick) light laminae. 50 to 80 % of the material is dolomicrite, the remaining fraction is anhydrite and calcite in varying amounts. The light, wider-spaced layers contain open structures filled by anhydrite (Fig. 42A). These interlaminar open structures clearly show the laminoid-fenestral LF-A fabric-type described by MÜLLER-JUNGBLUTH & TOSCHEK (1969). The dolomicrite surrounding the voids is arranged in tubular, worm-like patterns. At some places, bivalved ostracodes are embedded (Fig. 42B). Nodular structures of a completely different shape are also filled with anhydrite and are embedded in the more widely spaced light laminae in particular. The nodule surface is irregularly rounded and resembles cauliflower surface structure (Fig. 42C). In close proximity to these nodules the tubular dolomicrite is concentrically oriented and surrounds the nodules as a distinct halo (Fig. 42C).

Interpretation: The similarity of the nodular to biolaminoid facies type found in the Gavish Sabkha to the middle sequence of Plattendolomite deposits and modern biolaminated deposits demonstrated in Fig. 42 is so striking that the Gavish Sabkha may be used here as a model. The specific topography of the Gavish Sabkha is not meant here but rather the type of a coastal sabkha environment possessing sufficient water-saturated sediments to allow microbial communities to thrive. Several other examples of arid coastal environments exist, where similar structures can be found, e. g. at the Persian Gulf (PURSER, 1973; GOLUBIC & AWRAMIK, 1974; GOLUBIC & PARK, 1973; KINSMAN & PARK, 1976), in Mexico (JAVOR, 1979; MARGULIS et al., 1980) and the Shark Bay, Australia (LOGAN, 1961; GOLUBIC & HOFMANN, 1976; BAULD, 1984). The faint dark laminae may document microbial mats of the L_h-type which occurred sandwiched between the much more vertically extended L_v-laminae (Figs.

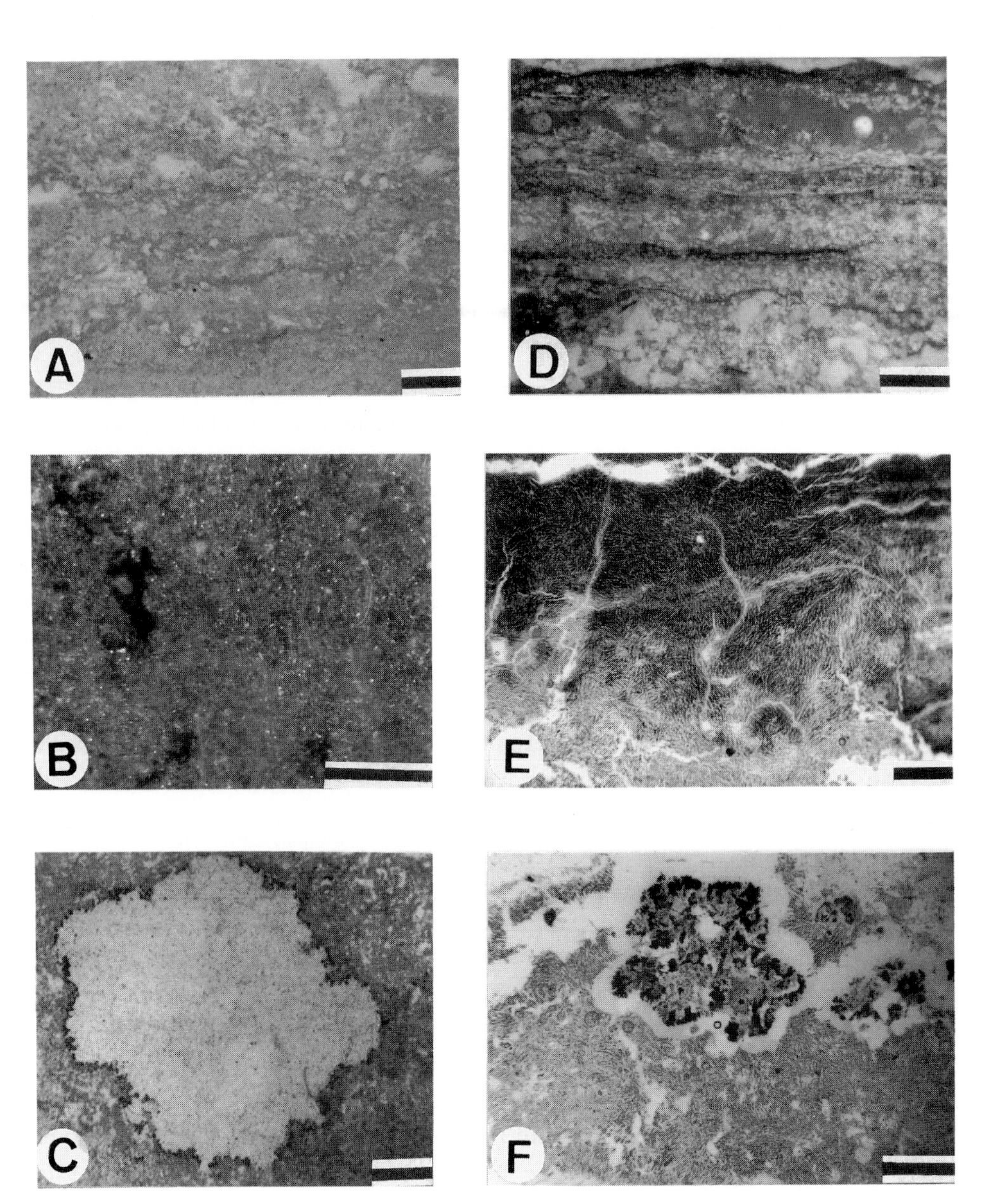

42A/42D), and the worm-shaped tubular dolomicrite may represent ghost structures of vertically or diagonally oriented filamentous organisms (Figs. 42B/42E). Evaporative pumping may have brought about a pre-

Fig. 42. Similarity of ancient and modern sabkha-type microstructures (A - C: Permian Zechstein Plattendolomite North Poland; D - F: Gavish Sabkha).

A) Interlayered faint dark and extended light laminae with open spaces oriented parallel to the dark laminae (laminoid fenestral LF-A fabric-type). Scale is 1 mm.

B) Tubular, worm-like patterns of dolomicrite surrounding the voids. Some bivalved ostracode shells being visible. Scale is 1 mm.

C) Anhydrite nodule surrounded by the worm-like dolomicrite. Scale is 1 mm.

D) Modern biolaminoid structures of the Gavish Sabkha (similar to A). Scale is 1 mm.

E) Tubular, worm-like patterns in modern L_V-laminae consist of filamentous cyanobacteria, coated with mg-calcite (similar to B). Scale is 1 mm.

F) Cauliflower-shaped nodule, made-up by Pleurocapsalean cyanobacteria, surrounded by tubular calcified filaments (Gavish Sabkha; similar to C). Scale is 2 mm.

---

enrichment of Mg over calcium which was bound already in early stages of diagenesis to the localized organic fraction of decaying cells. Aggregate-forming microorganisms with multiple fission, like the modern Pleurocapsalean which form the cauliflower nodules in the Gavish Sabkha, may have brought about the nodules of the Plattendolomite (Figs. 42C/42F). The anhydrite filling may be correlated to early (after death) or late diagenetic processes (replacement). In conclusion, it is proposed that the middle part of the PZ3 sequence developed in an intertidal to supratidal hypersaline coastal environment. It may have framed an adjacent platform margin where stromatolites of the stratiform type have developed more regular laminations.

3. 3. Lower Jurassic ironstone, Lorraine

3. 3. 1. Provenance of rock samples and previous work

The oolitic Minette ironstone of the northwestern part of Paris Basin (Toarcian - Aalenian, Luxembourg, Lorraine) is well described in terms of stratigraphy, sedimentology and geochemistry (BUBENICEK, 1971; THEIN, 1975; SIEHL & THEIN, 1978). The above authors suggested a depositional environment beneath the lower water line of a wide shelf area.

Fig. 43. Lower Jurassic ironstone, France: Microstructures.
A) An extended layer with ooids and oncoids (bottom) grades into a wavy laminated stromatolitic sequence of light and dark layers (top). Scale is 500 µm.
B) The ooid and oncoid-bearing layer shows some elongated fragments oriented parallel to the laminated sequence in A). Scale is 1 mm.
C) Dispersed and clustered arrangements of coated grains, the clustered being arranged in a lensoid pattern. Scale is 1 mm.
D) SEM-photography of a coated grain surrounded by a meshwork of filamentous microfossils. Scale is 100 µm.
E) Interior of oncoids showing meshworks of filamentous microfossils similar to those of the surrounding matrix. Scale is 100 µm.
F) Close-up of the interior showing a branched, fungal-like morphology of the microfossils. Scale is 3 µm.

The oolites were thought to be detritic particles derived from lateritic soils. The rock samples studied here were provided by K. DAHANAYAKE.

3. 3. 2. Microstructures

Description: Ooids and oncoids of the Minette ironstones consist of chamosite, siderite, magnetite and hematite. The groundmass is of calcite and limonite with some clay minerals admixed. In vertical succession, the ooid and oncoid bearing layers grade into wavy, alternating black and light laminae (Fig. 43A). The lower zone does not show these regular laminations but contains elongated fragments oriented parallel to the laminated sequence above (Fig. 43B). Oooids and oncoids are irregularly distributed within the lower zone. Some lie very close together and show a lensoid arrangement (Fig. 43C), others are dispersed within the widely spaced, light-colored matrix. Filamentous microfossils occur in abundance within the horizontally oriented laminae, the elongated fragments and the groundmass which carries the coated grains (Fig. 43D). The interior of the grains, when deeply etched, also shows filaments in abundance which are very similar to those of the surrounding matrix and the laminae (Fig. 43E). The morphology of the microfossils characterizes them as branched fungal-like microorganisms (Fig. 43F) which have formed mycelia (more condensed laminae) and looser meshes of hyphae.

Interpretation: The repetition of microfossils of the same taxonomical nature within different fabrics and the combination of horizontally

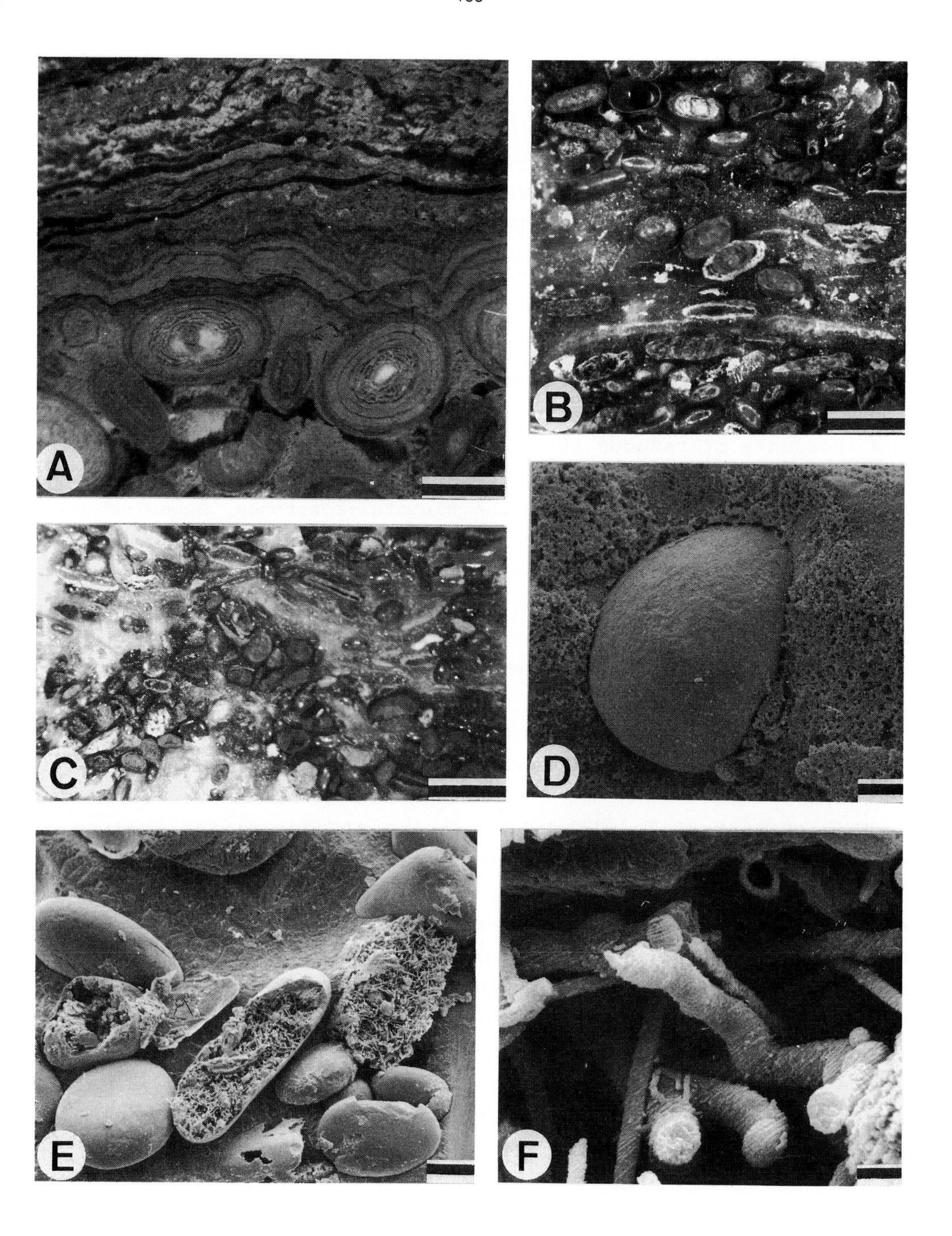

oriented laminae or laminae fragments and concentrically laminated coated grains of both oncoid and ooid characters indicate a microbial mat environment. It is reasonable to assume that the coated grains have been formed in situ and actually by the same microorganisms which have built the microbial mats. Fungi are known to be selective catalysts of metals and to enrich them in different redox states even within the same environment (TRUDINGER & SWAIN, 1979). The precipitation of ferric and ferrugenous iron may depend, however, on a microenvironmental differentiation created by the complexely interacting assimilative and dissimilative pathways which usually occur in organically enriched sediments.

Fungal mats may flourish in marine environments only if large amounts of organic matter are available and salinity is reduced. These conditions are found for example in estuaries and black-water lagoons. The abundance of reduced authigenic iron minerals (e. g. siderite), which is rare in normal marine sediments, additionally points to a brackish water environment. For the Minette ironstones, Bubenicek suggested a coastal tidal-flat region. Bituminous mudstones associated with the oolitic ironstone suggest that the environment was surrounded by rich vegetation. With respect to Upper Jurassic ironstones of the Swiss Jura, GYGI (1981) interpreted the depositional environment as a deep open marine shelf environment surrounded by tropical forests.

3. 4. Summary and conclusions

This chapter deals with some results of comparative studies on microbially mediated structures in rocks. Results of studies in modern stromatolite environments have been used in part to interpret the origin of microstructures. The question of biogenicity has been stressed with special reference to ooids occurring in rock specimens of Precambrian Gunflint iron formation and Lower Jurassic ironstone, Lorraine.

1. The following signatures indicate biogenicity and in-situ genesis of ooids and oncoids (Gunflint iron formation, Lorraine ironstone):

- Microfossils from the surrounding mat environment preserved in laminations of ooids,
- close associations of ooids and light and dark laminae which show wavy and microcolumnar morphologies and bear the same microfossils as in the laminations of the ooids,
- also close associations of ooids, intraclasts and lamina fragments,

- abundant enrichment of ooids within cavities, pockets and lenses (intralaminar unconformities typically occurring in abundance in microbial mats),
- the same mineral species within laminae of ooids and surrounding matrices.

2. The examples of Gunflint iron formation and Lorraine ironstones show that similar microstructures and minerals form under participation of different microbial phyla.
- In the case of Lower Jurassic ironstone/Lorraine, fungi have been the predominant frame builders.
- The filamentous microfossils in the Precambrian iron formation document a completely different type, often considered to be cyanobacteria, although a close morphological similarity exists also between the Gunflint filaments and modern iron bacteria (for further discussion, also of the limitations of drawing physiological inferences from form, see KNOLL & AWRAMIK, 1983).

3. In the light of studies on modern sabkha environments, microbial impact (with cyanobacteria as main frame-builders) is inferred considering the ghost structures of worm-shaped tubular dolomicrite, the nodular and biolaminoid structures and the mineralogy of the Permian Zechstein sequence. Numerous individual examinations have led to an understanding of carbonate precipitation in organic rich sediments such as microbial mats. KRUMBEIN (1974 a, b, 1979a) observed the precipitation of carbonates alongside sterilized cyanobacteria filaments which were added to isolates of marine chemoorganotrophic bacteria. Control experiments have shown that neither photosynthetic carbon dioxide fixation of cyanobacterial mats nor sterile experiments (autoclaved samples to kill all organisms) have brought about carbonate deposits. The results indicate the importance of chemoorganotrophic bacteria in the precipitation of carbonates during degradation of cyanobacterial organic matter. These experiments also revealed a positive correlation between Mg mole % in the calcite and in the applied medium. Furthermore, observations in natural habitats of microbial mats and in laboratory experiments indicate that Mg is preferentially bound to the bacterial biomass. BATHURST (1971) refers to the significance of the environment in this phenomenon (e. g. salinity, temperature, significance of certain organic products such as fatty acids, malate, NH_4). The environmental significance can be seen in the Gavish Sabkha, where

positive correlations between the inclination of the terrane and (1) salinity, (2) Mg-enrichment of the water, (3) layer thickness and productivity of cyanobacterial mats and (4) activity of chemoorganotrophs finally led to the accumulation of large amounts of high-magnesium calcite within stromatolitic growth structures.

4. PATHWAYS INVOLVED IN MICROBIAL SEDIMENT ACCRETION: A COMPLEMENTARY SUMMARY

The recovering of well-preserved biogenic structures in sedimentary rocks of early Proterozoic age, the reasoning that the kingdom of procaryotes, solely reigning at that time, had already developed metabolic traits including anoxygenic and oxygenic photosynthesis which we know from present-day forms, the conclusion that the early microbial systems mediated oxidation processes which finally led to the accumulation of free oxygen into the atmosphere, all this has received increasing attention in the autecology of present-day microorganisms.

In turn, a need was recognized for sedimentologists to understand the diversity and flexibility of microbial pathways in biological sediment accretion and strata-bound ore formation (MARGULIS & STOLZ, 1983). With a view to completing our records some of these aspects will be briefly summarized.

Basically, microbes interacting with a sedimentary environment use

1. an energy source which maintains a steady state equilibrium (thermodynamic stabilization against entropy),
2. an electron donor (reducing power) which regulates the energy flow through the cell and provides reduced carbon compounds for metabolic, morphologic and motility purposes,
3. nutrients (carbon, nitrogen, phosphorous and other compounds), crucial to build the necessary cell compounds,
4. a terminal electron acceptor for energy conversions and final steps of the oxidation-reduction reactions (KRUMBEIN, 1983).

The kingdoms of Monera and Protoctista (SCHWARTZ & MARGULIS, 1982) contain various different pathways to recover these sources. There are species which use light energy and others which use chemically bound energy, some recover electron donors from inorganic sources (for example water, H_2S) and others from organic sources. The carbon source can be CO_2 and in another case it is an organic compound (Table 11). Some species perform aerobic and others anaerobic respiration and fermentation. More than that, several groups including cyanobacteria are able to carry out alternatively photosynthesis and chemosynthesis, to switch from inorganic to organic electron donors and carbon sources (Table

11). This metabolic flexibility is an advantage in particular in an environment which is sometimes deprived of sufficient light, and where poisoning by hydrogen sulfide or oxygen occurs.

According to this great variety and flexibility of microbial metabolic pathways there is nearly no geochemical medium on Earth unacceptable to microorganisms. Consequently, mineralization processes induced or controlled by microorganisms are extremely heterogenous (LOWENSTAM, 1986).

TABLE 11. Microbial metabolic types forming biolaminated deposits (the processes of recovering energy, reducing power and nutrients are given the term "troph")

Energy source	Electron donor	Carbon source	Trophic types and examples
Light	Inorganic	Inorganic	Photo-lithoautotroph Cyanobacteria*
Light	Organic	Inorganic	Photo-organoautotroph Chloroflexaceae*
Light	Inorganic	Inorganic	Photo-lithoautotroph Chromatiaceae*
Light	Inorganic	Inorganic	Photo-lithoautotroph Chlorobiaceae
Light	Organic	Organic	Photo-organoheterotroph Rhodospirilliaceae*
Bond energy of inorganic compounds such as Fe^{2+}, S^{2-}, S^{o}, $S_2O_3{}^{2-}$	Inorganic	Inorganic	Chemo-lithoautotroph Iron and sulfur bacteria*
Bond energy of organic compounds	Organic	Organic	Chemo-organoheterotroph Sulfate-reducing bacteria + Fungi

* Switching from one source to another possible

Many different metabolic types are known to form mats (Table 11). In this volume, we have mentioned already the following types:

- cyanobacteria (photolithotrophs), frame-builders of mats in the Gavish Sabkha, the Solar Lake, the temperate humid siliciclastic systems and possibly in the PZ3-sequence,

- iron bacteria (chemolithotrophs), possibly involved in the Gunflint iron formation. Also microaerophilic sulfur bacteria, niched for example into gradients of deep sea hot vents, belong to this group,
- anaerobic filamentous sulfur-bacteria of the *Chloroflexus*-type (photo-organotrophs),
- fungi (chemo-organotrophs), stromatolite frame builders in the case of Lorraine ironstone.

Chemolithotrophs obtain energy via direct oxidation of reduced inorganic compounds such as

- S^{2-}, S^{o}, $S_2O_3{}^{2-}$ (e. g. *Thiotrix, Thiobacillus, Beggiatoa*). The biogeochemical products are sulfate and elemental sulfur. The latter product is alternatively deposited inside or outside the cells. It represents an intermediate product which becomes further oxidized;
- Fe^{2+} (e. g. *Leptothrix, Crenothrix*). Iron-oxidizing bacteria occur in abundance in lakes, streams and bogs of high latitudes. These organisms create a reducing environment in subsoils and subsurface sediments which facilitates the migration of ferrous iron. Electron-transfer reactions, catalyzed by these organisms, lead to changes in the oxidation state of iron in the extracellular environment: $Fe^{2+} - e^{-} \rightarrow Fe^{3+}$. Biogeochemical products are ferric oxide and hydroxide minerals. A connection between the distribution of these bacteria and the formation of ferromanganese nodules within dystrophic lakes and bogs is also suggested. In the early biosphere ferrous iron may have served abundantly as electron donor, the catalyzing organisms hereby producing a chemical environment with iron equivalent to a sulfuretum (HARTMAN, 1984). It is suggested that the huge iron deposits of the Precambrian era, the banded iron formations (BIF's), which contain most of the world's iron ore, generated by microorganisms transforming ferrous iron into the ferric state (LUNDGREN & DEAN, 1979).

Requirements of fungi (and chemo-organotrophic bacteria) are organic compounds. Many fungi excrete substances, called siderophores, into their external environment to transform Fe^{3+}-ions into chelate-complexes for transport into the cells. Also actinomycetes and myxobacteria possess this instrument.

The versatility of metabolic types involved in the formation and subsequent diversification of biolaminated deposits is well documented in terms of stromatolitic mineral assemblages which include

Heavy metal stromatolites (oxidic)
Heavy metal stromatolites (sulfidic)
Iron stromatolites
Carbonate stromatolites
Evaporitic stromatolites including Sabkha-type laminites
Phosphate stromatolites
Organic stromatolites (i.e. oil shales, oil, gas deposits)
Siliciclastic stromatolites

A microbial mat (the actualistic model of a stromatolite) considered as a functional association of different populations which are bond together by trophic chains (syntrophism), where metabolic products of the one species are the benefit for the others, raises immensely the geochemical potential of a microbially colonized substratum. For example, chemotrophic anaerobic sulfate-reducing bacteria (e. g. *Desulfovibrio*) occur in abundance in the black anaerobic strata of microbial mats. They produce quantities of hydrogen sulfide which are required by chemolithotrophic bacteria. Hydrogen sulfide on the other hand is known to react with soluble metals to produce insoluble metal sulfides (RENFRO, 1974). Iron-reducing bacteria use ferric ion from ferric hydroxide or from ferric oxides and also manganese from MnO_2 as an electron acceptor during anaerobic respiration. Thus they form large amounts of soluble ferrous ions in anaerobic environments (NEALSON, 1983a).

Important knowledge of different kinds of physiological biotransfer mechanisms of microbes is derived from technical assays. A growing interest is directed to the value of microbes being capable of mineral transfer and metal accumulation. Important technological efforts in this field involve microorganisms in biohydrometallurgical processes, in mineral concentrate leaching, in the clearance of waste water from heavy metals and other projects of engineering of microorganisms (see for example ROSSI & TORMA, 1983). The progress in these fields in turn is valuable to understand the geological potential of microbes in nature.

In terms of biomineralization processes, a distinction should be made between organisms which mineralize under controlled conditions and others which induce mineralization (LOWENSTAM, 1981, 1986; LOWENSTAM & WEINER, 1984). Biologically controlled mineralization indicates the importance of enzymes, but also proteins, carbohydrates and other materials that can bind, concentrate and thus enhance the oxidation of metal ions such as iron or manganese (NEALSON, 1983a, b).

Mat-forming and mat-colonizing microbes may mainly account for biologically induced mineralization processes which are well-defined by interactions between metabolite products, cations and anions present in the external environment. Thus the mineral types produced in a "pseudo-inorganic manner" (LOWENSTAM, 1986) may reflect much of the environment in which they assembled. For example, mg-calcite precipitation in the Gavish Sabkha is the result of high-rate primary production within the multilaminated mats, combined with bacterial break-down of organic carbon, generating oxidants and reductants and their interaction with metal ions supplied by seawater run-off or evaporation pumping. The example shows further that the relationship of carbonate to sulfate is largely regulated by the microbial activity.

Comparative experiments with isolates of living microbial mats and sterile organic matter (incubated to kill all living organisms) have shown that precipitation occurs especially well in the presence of living microbial communities (KRUMBEIN, 1974a, b, 1979a). DEXTER-DYER GROSOVSKY (1983), performing similar experiments with a solution of gold chloride, concluded that many cyanobacterial communities possess mechanisms for the active precipitation or flocculation of metallic gold. These experiments may shed light on live microbial involvement in the deposition of gold in the 2.4 billion-year-old Witwatersrand system in South Africa.

Weathering of original vulcanic metal deposits surrounding the sedimentary basins where microbial mats thrive is very important in cation supply. Other important sources of cations and anions brought into the vicinity of trapping and catalyzing microbial mats are hot vents (sulfides) and oceanographic upwelling systems both in higher and lower latitudes (phosphorites). Precambrian-Cambrian stromatolitic phosphorites have been reported by CHAURASIA (1984), which are considered to be the characteristic of a palaeoceanographic cold glacial upwelling system. Seaward winds support very often the penetration of upwelling waters rich in phosphorous and other compounds (e. g. opal from diatom tests) into shallow coastal lagoons. The lagoon's surface water becomes transported in offshore direction and replaced by the cold, nutrient-rich upwelling waters (REINECK & SINGH, 1980). Tertiary stromatolitic phosphorites formed by fungi have been reported by DAHANAYAKE & KRUM-

BEIN (1985). While phosphorites mediated by fungal stromatolites may have their base in organic-rich upwelling water, fungal mats catalyzing iron precipitation such as in the Lorraine ironstone may have been formed in warm shallow coastal systems supplied with nutrients from tropical rain forests.

Plant communities (typical xerophytes, rain forests, peat and swamp flores), animal communities (thickness of shells and other phenomena related to the solubility of carbonates in seawater), desert phenomena (evaporites, pseudomorphoses of soluble salts, carbonates, gypsum and opal and other mineral associations) and the often widely distributed "red beds" are repeatedly referred to in literature on paleoclimatology as useful tools. On the subject of stromatolites or microbial communities as paleoclimatological indicators there is, on the other hand, little discussion.

As a first step, the following climatological (paleoclimatological) model is proposed basing on individual types of stromatolites and their associated biomineral assemblages:

A) Ferriferous tropical (rainforest-mangrove related)
B) Phosphatic sub-tropical (upwelling related)
C) Carbonate sub-tropical (arid coast related)
D) Siliciclastic temperate (with iron sulfide/oxide sandwiching)
E) Ferric boreal (lacustrine bog ores with vivianite)
F) Ferrous boreal (lacustrine carbonate/sulfide sandwiching)

Many questions are still open on the chemical capacities of microbial communities (microbial mats), the sedimentary products of such microbial mats, their distribution in space and time (around the globe and through the geological periods) and the quantity of chemicals piled up and recycled during periods of time. Finding answers on these questions, we will eventually be able to piece together the picture of paleoclimates and climatic belts from the occurrence of major types of stromatolites as they are produced by major synergetic, ecological and climatic cross-relations and we will be able to set them into an actualistic and fossil strata framework as well as into a geographical framework. Further studies will help to decipher reaction changes and even traditions that may be derived from the primordial Earth or from the evolution of the bioplanet's dynamics.

Geochemistry is concerned with the concentration and distribution of chemical elements and their isotopes on Earth. Biogeochemistry, accordingly, is geochemistry controlled by the living and dead organic matter which constitutes the biosphere. Unlike the other main reservoirs (air, ocean and rocks or - in other terms - atmosphere, hydrosphere and lithosphere), the biosphere is not such a volume. It is spread and intertwined with the atmosphere, the hydrosphere and the lithosphere. Physiological biotransfer mechanisms of microbes have left their fingerprints in all these reservoirs. The progress in interdisciplinary work in the fields of biogeochemistry, geomicrobiology and paleomicrobiology will bring about more information to understand the geological potential of microbes in nature and may finally increase the attention in concepts which assert that the geochemistry of the exogenic system is regulated by life (LOVELOCK & MARGULIS, 1974; SCHWARTZ & MARGULIS, 1982; MARGULIS & STOLZ, 1983; KRUMBEIN, 1986a).

REFERENCES

ADAMS, J. E. & FRENZEL, H. N. (1950): Capitan barriere reef, Texas and New Mexico.- J. Geol. 58, 289-312.

AHARON, P., KOLODNY, Y. & SASS, E. (1977): Recent hot brine dolomitization in the Solar Lake, Gulf of Elat, Isotopic, chemical and mineralogical study.- J. Geol. 85, 27-48.

AITKEN, J. D. (1967): Classification and environmental significance of cryptalgal limestones and dolomites with illustrations from the Cambrian and Ordovician of Southwestern Alberta.- J. Sediment. Petrol. 37, 1163-1178.

AIZENSHTAT, Z., LIPINER, G. & COHEN, Y. (1984): Biogeochemistry of carbon and sulfur cycle in the microbial mats of the Solar Lake (Sinai).- In: COHEN, Y., CASTENHOLZ, R. W. & HALVORSON, H. O. (eds.): Microbial Mats: Stromatolites, 281-312, 498 p. (Alan Liss Publ., New York).

AWRAMIK, S. M. (1971): Precambrian columnar stromatolite diversity: Reflection of metazoan appearance.- Science 174, 825-827.

AWRAMIK, S. M. (1981): The Pre-Phanerozoic biosphere - three billion years of crisis and opportunities.- In: NITECKI, M. H. (ed.): Biotic crisis in ecological and evolutionary time, 83-102 (Academic Press, London).

AWRAMIK, S. M. & BARGHOORN, E. S. (1977): The Gunflint microbiota.- Precambrian Research 5, 121-142.

AWRAMIK, S. M., GEBELEIN, C. D. & CLOUD, P. (1978): Biogeologic relationships of ancient stromatolites and modern analogs.- In: KRUMBEIN, W. E. (ed.): Environmental biogeochemistry and geomicrobiology, vol. 1, 165-178 (Ann Arbor Science Publ. Inc., Ann Arbor, Michigan).

AWRAMIK, S. M., MARGULIS, L. & BARGHOORN, E. S. (1976): Evolutionary processes in the formation of stromatolites.- In: WALTER, M. R. (ed.): Developments in Sedimentology, vol. 20, 149-162 (Elsevier Scientific Publishing Company, Amsterdam).

AWRAMIK, S. M. & SEMIKHATOV, M. A. (1979): The relationship between morphology, microstructure, and microbiota in three vertically intergrading stromatolites from the Gunflint Iron Formation.- Canadian Journal of Earth Sciences 16, 484-495.

BARGHOORN, E. S. & TYLER, S. A. (1965): Microorganisms from the Gunflint Chert.- Science 147, 563-577.

BATHURST, R. G. C. (1971): Carbonate sediments and diagenesis.- Developments in Sedimentology 12 (Elsevier Scientific Publishing Company, Amsterdam).

BAULD, J. (1984): Microbial mat in marginal marine environments; Shark Bay, Western Australia and Spencer Gulf, South Australia.- In: COHEN, Y., CASTENHOLZ, R. W. & HALVORSON, H. O. (eds.): Microbial Mats: Stromatolites, 39-58, 498 p. (Alan Liss Publ., New York).

BERNER, R. A. (1980): Early diagenesis. A theoretical approach, 237 p. (Princeton University Press, Princeton).

BLACK, M. (1933): The algal sediments of Andros Island, Bahamas.- Phil. Trans. Roy. Soc. London B. 222, 165-192.

BLACKBURN, T. H. (1983): The microbial nitrogen cycle.- In: KRUMBEIN, W. E. (ed.): Microbial Geochemistry, 63-90, 330 p. (Blackwell Scientific Publications, Oxford).

BOON, J. J. (1984): Tracing the origin of chemical fossils in microbial mats: Biogeochemical investigations of Solar Lake cyanobacterial mats using analytical pyrolysis methods.- In: COHEN, Y., CASTENHOLZ, R. W. & HALVORSON, H. O. (eds.): Microbial Mats: Stromatolites, 313-342, 498 p. (Alan R. Liss, Inc., New York).

BOON, J. J., LEEUW, J. W. DE & KRUMBEIN, W. E. (1985): Biogeochemistry of Gavish Sabkha sediments. II. Pyrolysis mass spectrometry of the laminated microbial mat in the permanently water-covered zone before and after the desert sheetflood of 1979.- In: FRIEDMAN, G. M. & KRUMBEIN, W. E. (eds.): Hypersaline ecosystems - The Gavish Sabkha, vol. 53, 368-380, 484 p. (Springer, Heidelberg).

BRO LARSEN, E. (1936): Biologische Studien über die tunnelgrabenden Käfer auf Skallingen.- Videnskab Medd. Dansk Naturhist. Foren 100, 1-231.

BROCK, T. D. (1976): Biological techniques for the study of microbial mats and living stromatolites.- In: WALTER, M. R. (ed.): Stromatolites. Developments in Sedimentology, vol. 20, 21-30, 790 p. (Elsevier Publishing Company, Amsterdam).

BRODERICK, T. M. (1920): Economic geology and stratigraphy of the Gunflint iron district, Minnesota.- Economic Geology 15, 422-452.

BUBENICEK, L. (1971): Géologie du gisement de fer de Lorraine.- Bull. Centre Rech. Pau-SNPA 5, 223-320.

BUICK, R. (1984): Carbonaceous filaments from North Pole, Western Australia: Are they fossil bacteria in archaean stromatolites?- Precambrian Research 24, 157-172.

BUICK, R., DUNLOP, J. S. R. & GROVES, D. I. (1981): Stromatolite recognition in ancient rocks: an appraisal of irregularly laminated structures in an Early Archaean chert-barite unit from North Pole, Western Australia.- Alcheringa 5, 161-181.

BURST, J. F. (1965): Subaqueously formed shrinkage cracks in clay.- J. Sediment. Petrol. 35, 348-353.

CADEE, G. C. (1976): Sediment reworking of *Arenicola marina* on tidal flats in the Dutch Wadden Sea.- Netherlands Journal of Sea Research 10, 440-460.

CAMERON, B., DAMERON, D., JONES, J. R. (1985): Modern algal mats in intertidal and supratidal quartz sands, Northeastern Massachusetts, U.S.A.- In: CURRAN, H. A. (ed.): Biogenic structures: Their use in interpreting depositional environments, 211-223, 347 p. (SEPM, Tulsa, Oklahoma).

CARNEY, R. S. (1981): Bioturbation and biodeposition.- In: BOUCOT, A. J. (ed.): Principles of marine paleoecology, 357-400, 463 p. (Academic Press, London).

CASTENHOLZ, R. W. (1969): Thermophilic blue-green algae and the thermal environment.- Bacteriological Reviews 33, 476-504.

CASTENHOLZ, R. W. (1984): Composition of hot spring microbial mats: A summary.- In: COHEN, Y., CASTENHOLZ, R. W. & HALVORSON, H. O. (eds.): Microbial mats: Stromatolites, 101-120, 498 p. (Alan Liss Publ., New York).

CHAURASIA, P. K. (1984): A report on the glimpses of Precambrian-Cambrian stromatolitic phosphogenic Province of Karatan - phosphorite deposit, Kazakhstan, U.S.S.R.- Stromatolite Newsletter, vol. 11, 20-26.

CLOUD, P. E. (1965): Significance of the Gunflint (Precambrian) microflora.- Science 148, 27-45.

CLOUD, P. E. (1976): Beginnings of biospheric evolution and their biogeochemical consequences.- Paleobiology 2, 351-387.

CLOUD, P. E. & MORRISON, K. (1979): On microbial contaminants, micropseudofossils and the oldest records of life.- Precambrian Research 9, 81-91.

COHEN, Y. (1984): The Solar Lake cyanobacterial mats: Strategies of photosynthetic life under sulfide.- In: COHEN, Y., CASTENHOLZ, R. W. & HALVORSON, H. O. (eds.): Microbial mats: Stromatolites, 133-148, 498 p. (Alan Liss Publ., New York).

COHEN, Y., CASTENHOLZ, R. W. & HALVORSON, H. O. Eds. (1984): Microbial mats: Stromatolites, 498 p. (Alan Liss Publ., New York).

COHEN, Y., JORGENSEN, B. B., REVSBECH, N. P. & POPLAWSKI, R. (1986): Adaptation to hydrogen sulfide of oxygenic and anoxygenic photosynthesis among cyanobacteria.- Appl. Environ. Microbiol. 51, 398-407.

COHEN, Y., KRUMBEIN, W. E., GOLDBERG, M. & SHILO, M. (1977a): Solar Lake (Sinai). 1. Physical and chemical limnology.- Limnology and Oceanography 22, 597-608.

COHEN, Y., KRUMBEIN, W. E. & SHILO, M. (1977b): Solar Lake (Sinai). 2. Distribution of photosynthetic microorganisms and primary production.- Limnology and Oceanography 22, 609-620.

COHEN, Y., KRUMBEIN, W. E. & SHILO, M. (1977c): Solar Lake (Sinai). 3. Bacterial distribution and production.- Limnology and Oceanography 22, 621-634.

COLIJN, F. & KOEMAN, R. (1975): Das Mikrophytobenthos der Watten, Strände und Riffe um den Hohen Knechtsand in der Wesermündung.- Forschungsstelle für Insel- und Küstenschutz 26, 53-84.

DAHANAYAKE, K., GERDES, G. & KRUMBEIN, W. E. (1985): Stromatolites, oncolites and oolites biogenically formed in situ.- Die Naturwissenschaften 72, 513-518.

DAHANAYAKE, K. & KRUMBEIN, W. E. (1985): Ultrastructure of a microbial mat-generated phosphorite.- Miner. Depos. 20, 260-265.

DAHANAYAKE, K. & KRUMBEIN, W. E. (1986): Microbial structures in oolitic iron formations.- Miner. Depos. 21, 85-94.

DANIELLI, H. M. C. & EDINGTON, M. A. (1983): Bacterial calcification in limestone caves.- Geomicrobiology Journal 3, 1-16.

D'ARCY THOMPSON, (1984): On growth and form.- In: BONNER, J. T. (ed.): Reprint 1917, 345 p. (Cambridge Univ. Press, Cambridge)

DAUER, D. M., SIMON, J. L. (1976): Repopulation of the polychaete fauna of an intertidal habitat following natural defaunation: species equilibrium.- Oecologia 22, 99-117.

DAVIES, J. L. (1980): Geographical variation in coastal development.- In: CLAYTON, K. M. (ed.): Geomorphology Text 4, 212 p. (Longman Group Limited, London).

DAVIS, R. A. (1966): Willow River Dolomite: Ordovician analogue of moderna algal stromatolite environments.- J. Geol. 74, 908-923.

DAVIS, R. A. (1968): Algal stromatolites composed of quartz sandstone.- J. Sediment. Petrol. 38, 953-955.

FACHGRUPPE WASSERCHEMIE IN DER GES. DEUTSCHER CHEMIKER (Hrsg.) (1984): Deutsche Einheitsverfahren zur Wasser-, Abwasser- und Schlammuntersuchung (Chemie, Weinheim).

DEXTER-DYER, B., KRETZSCHMAR, M. & KRUMBEIN, W. E. (1984): Possible microbial pathways playing a role in the formation of Precambrian ore deposits.- J. Geol. Soc. 141, 251-262.

DEXTER-DYER-GROSOVSKY, B. (1983): Microbial role in Witwatersrand gold deposition.- In: WESTBROEK, P. & JONG, E. W. DE (eds.): Biomineralization and biological metal accumulation, 495-498, 533 p. (D. Reidel Publishing Company, Dordrecht).

DIMENTMAN, C. & SPIRA, Y. (1982): Predation of *Artemia* cysts by water tiger larva of the genus *Anacaena* (Coleoptera, Hydrophilidae).- Hydrobiologia 97, 163-165.

DUNLOP, J. S. R., MUIR, M. D., MILNE, V. A. & GROVES, D. I. (1978): A new microfossil assemblage from the Archaean of Western Australia.- Nature 274, 676-678.

ECCLESTON, M., KELLY, D. P. & WOOD, A. P. (1985): Autotrophic growth and iron oxidation and inhibition kinetics of *Leptospirillum ferrooxidans*.- In: CALDWELL, D. E., BRIERLEY, J. A. & BRIERLEY, C. L. (eds.): Planetary Ecology, 263-272, 591 p. (Van Nostrand Reinhold Company, New York).

ECKSTEIN, Y. (1970): Physicochemical limnology and geology of a meromictic pond on the Red Sea shore.- Limnology and Oceanography 15, 363-372.

EHRENBERG, C. G. (1839): Über das im Jahre 1686 in Curland vom Himmel gefallene Meteorpapier.- Annalen der Physik und Chemie 16, 187-188.

EHRLICH, A. (1978): The diatoms of the hypersaline Solar Lake (NE Sinai).- Israel Journal of Botany 27, 1-13.

EHRLICH, A., DOR, I. (1985): Photosynthetic microorganisms of the Gavish Sabkha.- In: FRIEDMAN, G. M. & KRUMBEIN, W. E. (eds.): Hypersaline Ecosystems - The Gavish Sabkha, vol. 53, 296-321, 484 p. (Springer, Berlin).

EISMA, D. (1980): Natural forces.- In: DIJKEMA, K. S., REINECK, H.-E. & WOLFF, W. J. (eds.): Geomorphology of the wadden sea area, vol. 1, 20-31 (Wadden Sea Working Group, Leiden).

EVANS, G. (1965): Intertidal flat sediments and their environments of deposition in the Wash.- Q. J. Geol. Soc. London 121, 209-245.

FABRICIUS, F. (1977): Origin of marine oöids and grapestones.- Contributions to Sedimentology 7, 1-77.

FERGUSON, J. & BURNE, R. V. (1981): Interactions between saline redbed groundwaters and peritidal carbonates, Spencer Gulf, South Australia: significance for models of stratiform copper ore genesis.- BMR Journal of Australian Geology & Geophysics 6, 3129-325.

FLÜGEL, E. (1982): Microfacies analysis of limestones, 633 p. (Springer, Berlin).

FRIEDMAN, G. M. (1965): A fossil shoreline reef in the Gulf of Elat (Aqaba).- Israel Journal of Earth-Sciences 14, 86-90.

FRIEDMAN, G. M. (1972): Significance of Red Sea in problem of evaporites and basinal limestones.- American Association of Petroleum Geologists Bulletin 56, 1072-1086.

FRIEDMAN, G. M. (1978): Solar Lake: A sea-marginal pond of the Red Sea (Gulf of Aqaba or Elat) in which algal mats generate carbonate particles and laminites.- In: KRUMBEIN, W. E. (ed.): Environmental Biogeochemistry and Geomicrobiology. The Aquatic Environment, vol. 1, 227-235, 394 p. (Ann Arbor Science Publ. Inc., Michigan).

FRIEDMAN, G. M., AMIEL, A. J., BRAUN, M. & MILLER, D. S. (1973): Generation of carbonate particles and laminites in algal mats - example from sea-marginal hypersaline pool, Gulf of Aqaba, Red Sea.- American Association of Petroleum Geologists Bulletin 57, 541-557.

FRIEDMAN, G. M. & GAVISH, E. (1971): Mediterranean and Red Sea (Gulf of Aqaba) beachrocks.- In: BRICKER, O. P. (ed.): Carbonate cements, 13-16, 376 p. (Johns Hopkins Univ. Press, Baltimore).

FRIEDMAN, G. M. & KRUMBEIN, W. E. Ed. (1985): Hypersaline ecosystems - The Gavish Sabkha. Ecological Studies vol. 53, 484 p. (Springer, Berlin).

FRIEDMAN, G. M. & SANDERS, J. E. (1978): Principles of sedimentology, 792 p. (John Wiley & Sons, New York).

FÜHRBÖTER, A., DETTE, H. H. & MANZENRIEDER, H. (1981): In-situ-Untersuchungen der Erosionsstabilität und der Durchlässigkeit von Wattböden. Bericht Nr. 506 des Leichtweiß-Instituts für Wasserbau der Technischen Universität Braunschweig (unveröffentlicht), 70 p. (TU, Braunschweig).

FÜHRBÖTER, A., DETTE, H. H., MANZENRIEDER, H. & NIESEL, S. (1983): In-situ-Untersuchungen der Erosionsstabilität und der Durchlässigkeit von Wattböden. - 2. Meßzeitraum 1982. Bericht Nr. 552 des Leichtweiß-Instituts für Wasserbau der Technischen Universität Braunschweig (unveröffentlicht).

GADOW, S. & REINECK, H.-E. (1969): Ablandiger Sandtransport bei Sturmfluten.- Senckenbergiana marit. 1, 63-78.

GALLAGHER, E. D., JUMARS, P. A. & TRUEBLOOD, D. D. (1983): Facilitation of soft-bottom benthic succession by tube builders.- Ecology 64, 1200-1216.

GARRELS, R. M., PERRY, E. A. & MACKENZIE, F. T. (1973): Genesis of Precambrian iron formations and the development of atmospheric oxygen.- Economic Geology 68, 1173-1179.

GARRETT, P. (1970): Phanerozoic stromatolites: Noncompetitive ecologic restriction by grazing and burrowing animals.- Science 169, 171-173.

GASIEWICZ, A. (1984): Eccentric ooids.- Neues Jahrbuch für Geologie und Paläontologie Monatshefte, 204-211.

GASIEWICZ, A., Gerdes, G., Krumbein, W. E. (1986): Platy dolomite (Permian) - A peritidal sabkha-type biolaminoid facies.- In preparation.

GAVISH, E. (1974): Mineralogy and geochemistry of a coastal sabkha near Nabek, Gulf of Elat.- In: GILL, D. (ed.): Abstracts of papers presented at the 1972/73 seminar of the Geol. Survey of Israel, 18-19.

GAVISH, E. (1975): Recent coastal sabkhas marginal to the gulfs of Suez and Elat, Red Sea.- Rapport Commision International Mer Mediterranée 23, 129-130.

GAVISH, E. (1980): Recent sabkhas marginal to the southern coasts of Sinai, Red Sea.- In: NISSENBAUM, A. (ed.): Hypersaline brines and evaporitic environments, 233-251, 270 p. (Elsevier Scientific Publishing Company, Amsterdam).

GAVISH, E., KRUMBEIN, W. E. & HALEVY, J. (1985): Geomorphology, mineralogy and groundwater geochemistry as factors of the hydrodynamic system of the Gavish Sabkha.- In: FRIEDMAN, G. M. & KRUMBEIN, W. E. (eds.): Hypersaline Ecosystems - The Gavish Sabkha, vol. 53, 186-217, 484 p. (Springer, Heidelberg).

GEBELEIN, C. D. (1976): The effects of the physical, chemical and biological evolution of the earth.- In: WALTER, M. R. (ed.): Stromatolites. Developments in Sedimentology, vol. 20, 499-515 (Elsevier Scientific Publishing Company, Amsterdam).

GEIKIE, S. A. (1905): The founders of geology, 486 p. (McMillan, New York).

GERDES, G. & HOLTKAMP, E. (1980): Sedimentologisch-biologische Kartierung der Wattengebiete von Mellum (südliche Nordsee).- Courier Forschungsinstitut Senckenberg 39, 1-185.

GERDES, G. & KRUMBEIN, W. E. (1985): Beobachtungen zur Lebensweise von *Pygospio elegans* (Spionidae, Polychaeta Sedentaria) im Farbstreifen-Sandwatt von Mellum.- Verhandlungen der Gesellschaft für Ökologie 13, 49-54.

GERDES, G. & KRUMBEIN, W. E. (1986): Potentielle silikoklastische Stromatolithe des unteren Supralitorals (südliche Nordsee) als begrenzende Faktoren gang- und röhrenbauender mariner Evertebraten.- Neues Jahrbuch für Geologie und Paläontologie Abhandlungen 172, 163-191.

GERDES, G., KRUMBEIN, W. E. & HOLTKAMP, E. M. (1985a): Salinity and water activity related zonation of microbial communities and potential stromatolites of the Gavish Sabkha.- In: FRIEDMAN, G. M. &

KRUMBEIN W. E. (eds.): Hypersaline Ecosystems - The Gavish Sabkha, vol. 53, 238-266, 484 p. (Springer, Heidelberg).

GERDES, G., KRUMBEIN, W. E. & REINECK, H.-E. (1982): Grenzgänger des Lebens - Ökologische Studien an zwei strandnahen Salzseen am Golf von Aqaba (Sinai).- Natur und Museum 112, 309-323.

GERDES, G., KRUMBEIN, W. E. & REINECK, H.-E. (1985b): The depositional record of sandy, versicolored tidal flats (Mellum Island, southern North Sea).- J. Sediment. Petrol. 55, 265-278.

GERDES, G., KRUMBEIN, W. E. & REINECK, H.-E. (1985c): Verbreitung und aktuogeologische Bedeutung mariner mikrobieller Matten im Gezeitenbereich der Nordsee.- Facies 12, 75-96.

GERDES, G., SPIRA, Y. & DIMENTMAN, C. (1985d): The fauna of the Gavish Sabkha and the Solar Lake - a comparative study.- In: FRIEDMAN, G. M. & KRUMBEIN, W. E. (eds.): Hypersaline Ecosystems - The Gavish Sabkha, vol. 53, 322-345, 484 p. (Springer, Berlin).

GERLACH, S. A. (1977): Attraction to decaying organisms as a possible cause for patchy distribution of nematodes in a Bermuda beach.- Ophelia 16, 151-165.

GIANI, D., GIANI, L., COHEN, Y. & KRUMBEIN, W. E. (1984): Methanogenesis in the hypersaline Solar Lake (Sinai).- FEMS Microbiol. Lett. 25, 219-224.

GIERE, O. (1975): Population structure, food relations and ecologic role of marine oligochaetes, with special reference to meiobenthic species.- Marine Biology 31, 139-156.

GIESENHAGEN, K. (1922): Über die systematische Deutung und die stratigraphische Stellung der ältesten Versteinerungen Europas und Nordamerikas mit besonderer Berücksichtigung der Cryptozoen und Oolithe. III. Teil: Über Oolithe.- Abhandlungen der Bayerischen Akademie der Wissenschaften Mathematisch-physikalische Klasse 24, 1-42.

GÖHREN, H. (1975): Zur Dynamik und Morphologie der holozänen Sandbänke im Wattenmeer zwischen Jade und Eider.- Die Küste 27, 28-49.

GOLUBIC, S. (1976): Organisms that build stromatolites.- In: WALTER, M. R. (ed.): Stromatolites, Developments in Sedimentology, vol. 20, 113-126, 790 p. (Elsevier Scientific Publishing Company, Amsterdam).

GOLUBIC, S. & AWRAMIK, S. M. (1974): Microbial comparison of stromatolite environments: Shark Bay, Persian Gulf and the Bahamas.- Geological Society of America Abstracts with Programs 6, 759-760.

GOLUBIC, S. & HOFMANN, H. J. (1976): Comparison of Holocene and mid-Precambrian Entophysalidaceae (Cyanophyta) in stromatolitic algal mats: cell division and degradation.- J. Paleontol. 50, 1074-1082.

GOLUBIC, S. & PARK, R. K. (1973): Biological dynamics of an algal mat - a sediment affecting microbial community (Persian Gulf).- In: Symposium on Environmental Biogeochemistry, 1973, p. 5 (Logan, Utah).

GOODWIN, A. M. (1956): Facies relations in the Gunflint Iron Formation.- Economic Geology 51, 565-595.

GOVETT, G. J. S. (1966): Origin of banded iron formations.- Geological Society of America Bulletin 77, 1191-1212.

GRAY, J. S. (1984): Ökologie mariner Sedimente (Übersetzung H. Rumohr), 1932 p. (Springer, Berlin).

GRESSLY, A. (1838): Observations géologiques sur le Jura Soleurois.- Neue Denkschr. allgem. schweizer. Ges. Naturwiss. 2.

GYGI, R. A. (1981): Oolitic iron formations: marine or not marine?.- Eclogae Geologicae Helvetiae 74, 233-254.

HAMBLIN, W. K. (1962): X-ray radiography in the study of structures in homogeneous sediments.- J. Sediment. Petrol. 32, 201-210.

HARTMAN, H. (1984): The evolution of photosynthesis and microbial mats; A speculation on the Banded Iron Formations.- In: COHEN, Y., CASTENHOLZ, R. W. & HALVORSON, H. O. (eds.): Microbial Mats: Stromatolites, 441-454, 498 p. (Alan Liss Publ., New York).

HAUSER, B. & MICHAELIS, H. (1975): Die Makrofauna der Watten, Strände, Riffe und Wracks um den Hohen Knechtsand in der Wesermündung.- Forschungsstelle für Insel- und Küstenschutz 26, 85-119.

HEIM, M. R. (1916): Monographie der Churfürsten-Mattstock-Gruppe. Teil 3: Lithogenese.- Beiträge zur geologischen Karte der Schweiz 20.

HEMPRICH, W. F. & EHRENBERG, C. G. (1828): Reisen in Aegypten, Libyen, Nubien und Dongala.- In: EHRENBERG, C. G. (ed.): Naturgeschichtliche Reisen durch Nord-Afrika und West-Asien in den Jahren 1820 bis 1825, (Mittler, Berlin).

HERTWECK, G. (1978): Die Bewohner des Wattenmeeres in ihren Auswirkungen auf das Sediment.- In: REINECK, H.-E. (ed.): Das Watt, 145-172, 185 p. (Kramer, Frankfurt am Main).

HOBSON, K. D. & GREEN, R. H. (1968): Asexual and sexual reproduction of *Pygospio elegans* in Barnstable Harbor, Mass.- Biol. Bull. 135, 410.

HOEK, C. VAN DEN, ADMIRAAL, W., COLIJN, F. & DE JONGE, N. N. (1979): The role of algae and seagrasses in the ecosystem of the Wadden Sea: A review.- In: WOLFF, W. J. (ed.): Flora and Vegetation of the Wadden Sea, vol. 3, 9-118 (Wadden Sea Working Group, Leiden).

HÖPNER, TH., ORLICZEK, CHR., GAUG, H. & KELLER, D. (1979): ASA-Anlage Oldenburg, ein universeller Analysenautomat für kolorimetrische Wasseranalysen.- Vom Wasser 52, 183-191.

HOFFMANN, C. (1942): Beiträge zur Vegetation des Farbstreifen-Sandwattes.- Kieler Meeresforschungen Sonderheft 4, 85-108.

HOFFMANN, C. (1949): Über die Durchlässigkeit dünner Sandschichten für Licht.- Planta 36, 48-56.

HOLTKAMP, E. (1985): The microbial mats of the Gavish Sabkha.- Thesis, 151 p., Universität Oldenburg.

HSÜ, K. J. & SIEGENTHALER, C. (1969): Preliminary experiments on the hydrodynamic movements induced by evaporation and their bearing on the dolomite problem.- Sedimentology 12, 11-25.

JACOBSEN, N. K. (1980): Form elements of the wadden sea area.- In: DIJKEMA, K. S., REINECK, H. E. & WOLFF, W. J. (eds.): Geomorphology of the wadden sea area, Rep. 1, 50-71 (Wadden Sea Working Group, Leiden).

JAMES, H. J. (1954): Sedimentary facies of iron-formations.- Economic Geology 49, 235-293.

JANNASCH, H. W. & WIRSEN, C. O. (1981): Morphological survey of microbial mats near deep-sea thermal vents.- Appl. Environ. Microbiol. 41, 528-538.

JAVOR, B. J. (1979): Ecology, physiology, and carbonate chemistry of blue-green algal mats, Laguna Guerrero Negro, Mexico (Dissertation, Eugene, Oregon).

JORGENSEN, B. B. (1977): The sulfur cycle of a coastal marine sediment (Limfjorden, Denmark).- Limnology and Oceanography 22, 814-832.

KALECSINSZKY, A. VON (1901): Über die Ungarischen warmen und heiszen Kochsalzseen als natürliche Wärmeakkumulatoren, sowie über die Herstellung von warmen Salzseen und Wärmeakkumulatoren.- Mathematisch naturwissenschaftliche Berichte Ungarn 19, 51-54.

KALKOWSKY, E. (1908): Oolith und Stromatolith im norddeutschen Buntsandstein.- Zeitschrift der Deutschen Geologischen Gesellschaft 60, 68-125.

KAZMIERCZAK, J. & KRUMBEIN, W. E. (1983): Identification of calcified coccoid cyanobacteria forming stromatoporoid stromatolites.- Lethaia 16, 207-213.

KENDALL, A. C. (1979): Continental and supratidal (Sabkha) evaporites.- In: WALKER, R. G. (ed.): Facies Models, 159-174.

KINSMAN, D. J. J. & PARK R. K. (1976): Algal belt and coastal sabkha evolution, Trucial Coast, Persian Gulf.- In: WALTER, M. R. (ed.): Stromatolites, 421-433 (Elsevier Scientific Publishing Company, Amsterdam).

KITANO, Y., KANAMORI, N. & TOKUYAMA A. (1969): Effects of organic matter on solubilities and crystal form of carbonates.- Amer. Zool. 9, 681-688.

KNOLL, A. H. (1985a): A paleobiological perspective on sabkhas.- In: FRIEDMAN, G. M. & KRUMBEIN, W. E. (eds.): Hypersaline Ecosystems - The Gavish Sabkha, 407-425, 484 p. (Springer, Berlin).

KNOLL, A. H. (1985b): Patterns of evolution in the Archean and Proterozoic eons.- Paleobiology 11, 53-64.

KNOLL, A. H. (1985c): The distribution and evolution of microbial life in the late Proterozoic era.- Ann. Rev. Microbiol. 39, 391-417.

KNOLL, A. H. & AWRAMIK, S. M. (1983): Ancient microbial ecosystems.- In: KRUMBEIN, W. E. (ed.): Microbial Geochemistry, 287-315, 330 p. (Blackwell Scientific Publications, Oxford).

KNOLL, A. H., BARGHOORN, E. S. & AWRAMIK, S. M. (1978): New microorganisms from the Aphebian Gunflint Iron Formation, Ontario.- J. Paleontol. 52, 976-992.

KRETZSCHMAR, M. (1982): Fossile Pilze in Eisen-Stromatolithen von Warstein (Rheinisches Schiefergebirge).- Facies 7, 237-260.

KRUMBEIN, W. E. (1974a): Mikrobiologische Untersuchungen zur Fällung von Kalziumkarbonat aus Meerwasser.- In: BIOLOGISCHE ANSTALT HELGOLAND: Jahresbericht 1973, 50-54, 104 p. (Boysen & Co, Heide).

KRUMBEIN, W. E. (1974b): On the precipitation of aragonite on the surface of marine bacteria.- Die Naturwissenschaften 61, 167-167.

KRUMBEIN, W. E. (1978): Algal mats and their lithification.- In: KRUMBEIN, W. E. (ed.): Environmental Biogeochemistry and Geomicrobiology. The Aquatic Environment, vol. 1, 209-225, 394 p. (Ann Arbor Science Publ. Inc., Michigan).

KRUMBEIN, W. E. (1979a): Calcification by bacteria and algae.- In: TRUDINGER, P. A. & SWAINE, D. J. (eds.): Biogeochemical Cycling of Mineral-forming Elements, vol. 3, 47-68, 612 p. (Elsevier Scientific Publishing Company, Amsterdam).

KRUMBEIN, W. E. (1979b): Photolithotrophic and chemoorganotrophic activity of bacteria and algae as related to beachrock formation and degradation (Gulf of Aqaba, Sinai).- Geomicrobiology Journal 1, 139-203.

KRUMBEIN, W. E. (1979c): Über die Zuordnung der Cyanophyten.- In: KRUMBEIN, W. E. (ed.): Cyanobakterien - Bakterien oder Algen?, 33-48, 130 p. (Universität, Oldenburg).

KRUMBEIN, W. E. (1983): Stromatolites - The challenge of a term in space and time.- Precambrian Research 20, 493-531.

KRUMBEIN, W. E. (1986a): Biotransfer of minerals by microbes and microbial mats.- In: LEADBEATER, B. S. C. & RIDING, R. (eds.): Biomineralization in Lower Plants and Animals, 55-72 (Oxford University Press, Oxford).

KRUMBEIN, W. E. (1986b): Die Entdeckung inselbildender Mikroorganismen.- In: GERDES, G., KRUMBEIN, W. E. & REINECK, H.-E. (eds.): Mellum - Portrait einer Insel (Kramer, Frankfurt am Main).

KRUMBEIN, W. E., BUCHHOLZ, H., FRANKE, P., GIANI, D., GIELE, C. & WONNEBERGER, C. (1979): O_2 and H_2S coexistence in stromatolites. A model for the origin of mineralogical lamination in stromatolites and banded iron formations.- Die Naturwissenschaften 66, 381-389.

KRUMBEIN, W. E. & COHEN, Y. (1974): Biogene, klastische und evaporitische Sedimentation in einem mesothermen monomiktischen ufernahen See (Golf von Aqaba).- Geologische Rundschau 63, 1035-1065.

KRUMBEIN, W. E. & COHEN, Y. (1977): Primary production, mat formation and lithification: Contribution of oxygenic and facultative anoxygenic cyanobacteria.- In: FLÜGEL, E. (ed.): Fossil Algae, 37-56, 375 p. (Springer, Berlin).

KRUMBEIN, W. E., COHEN, Y. & SHILO, M. (1977): Solar Lake (Sinai) 4. Stromatolitic cyanobacterial mats.- Limnology and Oceanography 22, 635-656.

LITTLE-GADOW, S. (1978): Sedimente und Chemismus.- In: REINECK, H.-E. (ed.): Das Watt, Ablagerungs- und Lebensraum, 51-62, 185 p. (Kramer, Frankfurt am Main).

LOGAN, B. W., REZAK, R. & GINSBURG, R. N. (1964): Classification and environmental significance of algal stromatolites.- J. Geol. 72, 68-83.

LOGAN, P. W. (1961): Cryptozoan and associate stromatolites from the Recent of Shark Bay, Western Australia.- J. Geol. 69, 517-533.

LOVELOCK, J. E. & MARGULIS, L. (1974): Atmospheric homeostasis by and for the biosphere: the Gaia hypothesis.- Tellus 26, 2-9.

LOWENSTAM, H. A. (1981): Minerals formed by organisms.- Science 211, 1126-1131.

LOWENSTAM, H. A. (1986): Mineralization processes in monerans and protoctists. - In: LEADBEATER, S. C. & RIDING, R. (Eds.): Biomineralization in lower plants and animals, 1-17, 401 p. (Systematics Association, Clarendon Press, Oxford).

LOWENSTAM, H. A. & WEINER, S. (1983): Mineralization by organisms and the evolution of biomineralization.- In: WESTBROEK, P. & JONG, E. W. DE (eds.): Biomineralization and biological metal accumulation, 191-204, 533 p. (D. Reidel Publishing Company, Dordrecht).

LUCAS, J. & PREVOT, L. (1984): Synthèse de l'apatite par voie bactérienne à partir de matière organique phosphatée et de divers carbonate de calcium dans des eaux douces et marines naturelles.- Chemical Geology 42, 101-118.

LUDWIG, R. & THEOBALD, G. (1852): Über die Mitwirkung der Pflanzen bei der Ablagerung des kohlensauren Kalkes.- Annalen der Physik und Chemie 87, 91-107.

LUNDGREN, D. G. & DEAN, W. (1979): Biogeochemistry of iron.- In: TRUDINGER, P. A. & SWAINE, D. J. (eds.): Biogeochemical Cycling of Mineral-Forming Elements, 211-251, 612 p. (Elsevier Publishing Company, Amsterdam).

LUSERKE, M. (1957): Juist - Verwunschene Insel.- Merian 3X, 15-20.

MACKENZIE, D. B. (1968): Studies for students: Sedimentary features of Alemeda Avenue cut, Denver, Colorado.- The Mountain Geologist 5, 3-14.

MACKENZIE, D. B. (1972): Tidal sand deposits in lower Cretaceous Dakota Group near Denver, Colorado.- The Mountain Geologist 9, 269-277.

MANZENRIEDER, H. (1984): Die biologische Verfestigung von Wattflächen aus der Sicht des Ingenieurs.- Veröffentlichungen der Naturforschenden Gesellschaft zu Emden 1984, 1-60.

MARGULIS, L., ASHENDORF, D., BANERJEE, S., FRANCIS, S., GIOVANNONI, S., STOLZ, J. F., BARGHOORN, E. S. & CHASE, O. (1980): The microbial community in the layered sediment at Laguna Figueroa, Baja California, Mexico: does it have Precambrian analogues?.- Precambrian Research 11, 93-123.

MARGULIS, L. & SCHWARTZ, K. V. (1982): Five kingdoms: Guide to the phyla of life on Earth, 338 p. (W. H. Freeman and Company, San Francisco).

MARGULIS, L., STOLZ, J. (1983): Microbial systematics and a Gaian view of the sediments.- In: WESTBROEK, P. & JONG, E. W. DE (eds.): Biomineralization and Biological Metal Accumulation, 27-53, 533 p. (Reidel Publ. Comp., Dordrecht).

MARKUN, C. D. & RANDAZZO, A. F. (1980): Sedimentary structures in the Gunflint Iron Formation, Schreiber Beach, Ontario.- Precambrian Research 12, 287-310.

MARTIN, T. C. & WYATT, J. T. (1974): Extracellular investments in blue-green algae with particular emphasis on the genus *Nostoc*.- Journal of Phycology 10, 204-210.

MASSARI, F. (1980): Cryptalgal fabrics in the Rosso Ammonitico sequences of Venetian Alps Roma.

MEYER, H. & MICHAELIS, H. (1980): Das Makrobenthos des westlichen "Hohe Weges".- Forschungsstelle für Insel- und Küstenschutz 31, 91-156.

MITTERER, R. M. (1971): Influence of natural organic matter on $CaCO_3$ precipitation.- In: BRICKER, O. C. (ed.): Carbonate Cements, 252-258, 376 p. (Hopkins, Baltimore).

MITTERER, R. M. (1972): Biogeochemistry of aragonite mud and oolites.- Geochimica et Cosmochimica Acta 36, 1407-1412.

MONTY, C. L. V. (1973): Precambrian background and Phanerozoic history of stromatolitic communities, an overview.- Annales de la Societé Géologique de Belgique 96, 585-624.

MONTY, C. L. V. (1976): The origin and development of cryptalgal fabrics.- In: WALTER, M. R. (ed.): Stromatolites, 193-249, 790 p. (Elsevier Scientific Publishing Company, Amsterdam).

MONTY, C. L. V. (1977): Evolving concepts on the nature and the ecological significance of stromatolites.- In: FLÜGEL, E. (ed.): Fossil Algae, 15-35, 375 p. (Springer, Berlin).

MÜLLER-JUNGBLUTH, W. V. & TOSCHEK, P. H. (1969): Karbonatsedimentologische Arbeitsgrundlagen.- Science 4, 1-32.

MÜLLER, O. F. (1777): Flora Danica vol. IV, Kopenhagen.

NEALSON, K. H. (1983): The microbial iron cycle.- In: KRUMBEIN, W. E. (ed.): Microbial Geochemistry, 159-190, 330 p. (Blackwell Scientific Publications, Oxford).

NEALSON, K. H. (1983): The microbial manganese cycle.- In: KRUMBEIN, W. E. (ed.): Microbial Geochemistry, 191-222, 330 p. (Blackwell Scientific Publications, Oxford).

NEUMANN, J. (1968): Solar Lakes and solar energy.- Nature 219, 851-852.

NOVITSKY, J. A. (1983): Apparent microbial precipitation of calcium carbonate in seawater: Importance of pH.- In: HALLBERG, R. O. (ed.): Environmental Biogeochemistry, vol. 35, 259-265, 576 p. (Ecological Bulletins Publishing House, Stockholm).

OEHLER, J. H. (1972): "Stromatoloids" from Yellowstone Park, Wyoming.- Geological Society of America Abstracts with Programs 4, 212-213.

OERSTEDT, O. S. (1841): Beretning om en Excursion til Trindelen, en Alluvialdannelse i Odensefjord.- Naturhistorisk Tidskrift 3, 552-569.

PANG, P. C. & NRIAGU, J. O. (1976): Distribution and isotope composition of nitrogen in Bay of Quinte (Lake Ontario) sediments.- Chem. Geol. 18, 93-105.

PARK, R. (1976): A note on the significance of lamination in stromatolites.- Sedimentology 23,379-393.

PERYT, T. M., CZAPOWSKI, G., DEBSKI, J. & PIZON, A. (1985): Model of sedimentation of Zechstein evaporites at the Leba Elevation.- Przeglad Geol. 33, 204-211.

PETTIJOHN, E. J. (1975): Sedimentary rocks, 628 p. (Harper & Row, New York).

PETTIJOHN, F. J. & POTTER, P. E. (1964): Atlas and glossary of primary sedimentary structures, 370 p. (Springer, Berlin).

PIA, J. (1927): Thallophyta.- In: HIRMER, M. (ed.): Handbuch der Paläobotanik, vol. 1, 1-136.

POR, F. D. (1968): Solar Lake on the shores of the Red Sea.- Nature 218, 860-861.

POR, F. D. (1969): Limnology of the heliothermal Solar Lake on the coast of Sinai (Gulf of Elat).- Verhandlungen Internationale Vereinigung Limnologie 17, 1031-1034.

POR, F. D. (1975): The Coleoptera-dominated fauna of the hypersaline Solar Lake (Gulf of Elat, Red Sea).- Tenth European Symposium on Marine Biology 2, 563-573.

POTTS, M., KRUMBEIN, W. E. & METZGER, J. (1978): Nitrogen fixation rates in anaerobic sediments determined by acetylene reduction, a new 15N field assay, and simultaneous total N15N determination.- In: KRUMBEIN, W. E. (ed.): Environmental Biogeochemistry and Geomicrobiology, vol. 3, 753-769, 711 p. (Ann Arbor Science Publ. Inc., Michigan).

PRATT, B. R. (1982): Stromatolite decline - a reconsideration.- Geology 10, 512-515.

PURSER, B. H. (1980): Les paléosabkhas du Miocène inf. dans le SE de l'Iran.- Bull. Centre Rech. Pau-SNPA 4, 235-244.

PURSER, B. H. (1985): Coastal evaporite systems.- In: FRIEDMAN, G. M. & KRUMBEIN, W. E. (eds.): Hypersaline Ecosystems -The Gavish Sabkha, 72-102, 478 p. (Springer, Berlin).

PURSER, B. H. (Ed.) (1973): The Persian Gulf, 471 p. (Springer, Berlin).

RANWELL, D. S. (1972): Ecology of salt marshes and sand dunes, 258 p. (Chapman & Hall, London).

RASMUSSEN, E. (1953): Asexual reproduction in *Pygospio elegans* Claparede (Polychaeta Sedentaria).- Nature 171, 1161-1162.

RASMUSSEN, E. (1973): Systematics and ecology of the Isefjord marine fauna (Denmark).- Ophelia 11, 1-495.

REINECK, H.-E. (1963): Sedimentgefüge im Bereich der südlichen Nordsee.- Abhandlungen der Senckenbergischen Naturforschenden Gesellschaft 505, 1-138.

REINECK, H.-E. (1967): Parameter von Schichtung und Bioturbation.- Geologische Rundschau 56, 420-438.

REINECK, H.-E. (1970): Reliefguß und projizierbarer Dickschliff.- Senckenbergiana maritima 2, 61-66.

REINECK, H.-E. (1976): Drift ice action on tidal flats, North Sea.- Rev. Geogr. Montreal 30, 197-200.

REINECK, H.-E. (1979): Rezente und fossile Algenmatten und Wurzelhorizonte.- Natur und Museum 109, 290-296.

REINECK, H.-E. & GERDES, G. (1984): Auf der Suche nach Grenzgängern: Eine Reise nach Neuwerk-Scharhörn.- Natur und Museum 114, 305-312.

REINECK, H.-E. & SINGH, I. B. (1980): Depositional sedimentary environments, 549 p. (Springer, Berlin).

REINKE, J. (1903): Die Entwicklungsgeschichte der Dünen an der Westküste von Schleswig.- Sitzungsberichte der Königlich Preussischen Akademie der Wissenschaften 1903, 281-295.

REISE, K. (1985): Tidal flat ecology, 191 p. (Springer, Berlin).

REMANE, A. (1940): Einführung in die zoologische Ökologie der Nord- und Ostsee.- In: GRIMPE, G. & WAGNER, E. (eds.): Die Tierwelt der Nord- und Ostsee, Becker und Erler.

RENFRO, A. R. (1974): Genesis of evaporite-associated stratiform metalliferous deposits - A sabkha process.- Economic Geology 69, 33-45.

RICHTER, D. K., HERFORTH, A. & OTT, E. (1979): Pleistozäne, brackische Blaugrünalgenriffe mit Rivularia haematites auf der Perachorahalbinsel bei Korinth (Griechenland).- Neues Jahrbuch für Geologie und Paläontologie Abhandlungen 159, 14-40.

RICHTER, R. (1926): Eine geologische Exkursion in das Wattenmeer.- Natur und Museum 56, 289-307.

RIPPKA, R., DERUELLES, J., WATERBURY, J. B., HERDMAN, M. & STANIER, R. Y. (1979): Generic assignments, strain histories and properties of pure cultures of cyanobacteria.- J. Gen. Microbiol. 111, 1-61.

ROSSL, G. & TORMA, A. E. (1983): Recent progress in biohydrometallurgy, 752 p. (Associazione Mineraria Sarda, Iglesias).

ROTHPLETZ, A. (1892): Ueber die Bildung der Oolithe.- Botanisches Zentralblatt 51, 265-268.

SCHÄFER, W. (1972): Ecology and palaeoecology of marine environments, 568 p. Edinburgh.

SCHOPF, J. W. & WALTER, M. R. (1980): Archaean microfossils and "microfossil-like" objects - a critical appraisal.- In: GLOVER, J. E. & GROVES, D. J. (eds.): Extended Abstracts, Second Internat. Archaean Symp. Pertz, Aust., 23-24 (Geol. Soc. Aust. & Internat. Geol. Correlation Pro, Perth).

SCHULZ, E. (1936): Das Farbstreifensandwatt und seine Fauna, eine ökologisch-biozönotische Untersuchung an der Nordsee.- Kieler Meeresforsch. 1, 359-378.

SCHULZ, E. & MEYER, H. (1940): Weitere Untersuchungen über das Farbstreifensandwatt.- Kieler Meeresforschungen Sonderheft 4, 321-336.

SCHWARZ, A. (1936): Der Lichteinfluß auf die Fortbewegung, die Einregelung und das Wachstum bei einigen niederen Tieren (*Littorina, Cardium, Mytilus, Balanus, Teredo, Sabellaria*).- Senckenbergiana 14, 429-454.

SCHWARZ, H.-U., EINSELE, G. & HERM, D. (1975): Quartz-sandy, grazing-contoured stromatolites from coastal embayments of Mauritania, West Africa.- Sedimentology 22, 539-561.

SEILACHER, A. (1951): Der Röhrenbau von *Lanice conchilega* (Polychaeta). Ein Beitrag zur Deutung fossiler Lebensspuren.- Senckenbergiana 32, 267-280.

SEILACHER, A. (1953): Studien zur Palichnologie I. Über die Methoden der Palichnologie.- Neues Jahrbuch für Geologie und Paläontologie Abhandlungen 96, 421-452.

SHEARMAN, D. J. (1978): Evaporites of coastal sabkhas.- In: DEAN, W. E. & SCHREIBER, B. C. (eds.): Marine evaporites, Soc. Econ. Paleontologists and Mineralogists Notes No. 4, 6-42.

SHINN, E. A. (1969): Submarine lithification of Holocene carbonate sediments in the Persian Gulf.- Sedimentology 12, 109-144.

SHINN, E. A. (1983): Tidal flat environment.- In: SCHOLLE, P. A., BEBOUT, D. G. & MOORE, C. H. (eds.): Carbonate depositional environments, AAPG Mem., vol. 33, 172-210, 708 p. Tulsa.

SIEHL, A. & THEIN, J. (1978): Geochemische Trends in der Minette (Jura, Luxembourg/Lothringen).- Geologische Rundschau 67, 10552-1077.

SIMONE, L. (1981): Ooids: A Review.- Earth-Science Reviews 16, 319-355.

SKYRING, G. W., CHAMBRS, L. A. & BAULD, J. (1983): Sulfate reduction in sediments colonized by cyanobacteria, Spencer Gulf, South Australia.- Australian Journal of Marine and Freshwater Research 34, 359-374.

SMIDT, E. (1951): Animal production in the Danish Wadden Sea.- Medd. Komm. Danmarks Fiskeri og Havundersogelser 11, 1-151.

SNEH, A. & FRIEDMAN, G. M. (1985): Hypersaline sea-marginal flats of the gulfs of Elat and Suez.- In: FRIEDMAN, G. M. & KRUMBEIN, W. E. (eds.): Hypersaline Ecosystems - The Gavish Sabkha, 103-135, 484 p. (Springer, Berlin).

STAL, L. (1985): Nitrogen-fixing cyanobacteria in a marine microbial mat (Dissertation, Rijksuniversiteit Groningen).

STAL, L. J. & GEMERDEN, H. VAN (1984): Microbielle Matten.- Natur en Techniek 11, 871-889.

STAL, L. J., GEMERDEN, H. VAN & KRUMBEIN, W. E. (1984a): The simultaneous assay of chlorophyll and bacteriochlorophyll in natural microbial communities.- Journal of Bacteriological Methods 2, 295-306.

STAL, L. J., GEMERDEN, H. VAN & KRUMBEIN, W. E. (1985): Structure and development of a benthic marine microbial mat.- FEMS Microbiology Ecology 31, 111-125.

STAL, L. J., GROßBERGER, S. & KRUMBEIN, W. E. (1984b): Nitrogen fixation associated with the cyanobacterial mat of a marine laminated microbial ecosystem.- Marine Biology 82, 217-224.

STAL, L. J. & KRUMBEIN, W. E. (1981): Aerobic nitrogen fixation in pure cultures of a benthic marine *Oscillatoria* (cyanobacteria).- FEMS Microbiol. Lett. 11, 295-298.

STAL, L. J. & KRUMBEIN, W. E. (1985): Oxygen protection of nitrogenase in the aerobically nitrogen fixing, non-heterocystous cyanobacterium *Oscillatoria* sp.- Arch. Microbiol. 143, 72-76.

STAL, L. J., KRUMBEIN, W. E. & GEMERDEN, H. VAN (1984c): Das Farbstreifen-Sandwatt - Ein laminiertes mikrobielles Ökosystem im Wattenmeer.- Veröffentlichungen der Naturforschenden Gesellschaft zu Emden 7, 1-60.

STANIER, R. Y. & COHEN-BAZIRE, G. (1977): Phototrophic prokaryotes: the cyanobacteria.- Ann. Rev. Microbiol. 31, 225-274.

STENSEN, N. (1967): Das Feste im Festen. Vorläufer einer Abhandlung über Festes, das in der Natur in anderem Festen eingeschlossen ist (Florenz 1669).- In: BALKE, S., GERIKE, H., HARTNER, W., KERSTEIN, G. & KLEMM, F. et. al (eds.): Ostwalds Klassiker der exakten Wissenschaften Neue Folge, vol. 3, 8-255 (Akademische Verlagsgesellschaft, Frankfurt am Main).

STOLZ, J. F. (1983): Fine structure of the stratified microbial community at Laguna Figueroa, Baja California, Mexico. I. Methods of in situ study of the laminated sediments.- Precambrian Research 20, 479-492.

STREIF, H., KÖSTER, R. (1978): Zur Geologie der deutschen Nordseeküste.- Die Küste 32, 30-49.

STROTHER, P. K., KNOLL, A. H. & BARGHOORN, E. S. (1983): Microorganisms from the Late Precambrian Narssarssuk Formation, North-Western Greenland.- Palaeontol. 26, 1-32.

TEBBUTT, G. E., CONLEY, C. D. & BOYD, D. W. (1965): Lithogenesis of a distinctive carbonate rock fabric.- Wyoming University Contributions Geology 4, 1-13.

TEICHERT, C. (1958): Concepts of facies.- Bull. Amer. Assoc. Petrol. Geol. 42, 2718-2744.

TEICHERT, C. (1970): Oolite, oolith and ooid: Discussion.- Bull. Amer. Ass. Petrol. Geol., 54, 26.

THEIN, J. (1975): Sedimentologisch-stratigraphische Untersuchungen in der Minette des Differdinger Beckens (Luxemburg). Publ. Serv. Geol. Luxembourg, 24, 60 S.

TISLJAR, J. (1983): Coated grain facies in the Lower Cretaceous of the Outer Dinarides (Yugoslavia).- In: PERYT, T. M. (ed.): Coated Grains, 566-575, 655 p. (Springer, Berlin).

TRUDINGER, P. A. & SWAINE, D. J. (Eds.) (1979): Biogeochemical cycling of mineral-forming elements, 612 p. (Elsevier Scientific Publishing Company, Amsterdam).

TRUSHEIM, F. (1935): Eine Titaneisenerz-Seife von Wangerooge.- Senckenbergiana 17, 62-72.

TYLER, S. A. & TWENHOFEL, W. H. (1952): Sedimentation and stratigraphy of the Huronian of Upper Michigan.- Amer. J. Sci. 250, 1-27.

WALTHER, J. (1885): Die gesteinsbildenden Kalkalgen des Golfes von Neapel und die Entstehung strukturloser Kalke.- Zeitschrift der Deutschen Geologischen Gesellschaft 37, 329-357.

WALTHER, J. (1888): Die Korallenriffe der Sinaihalbinsel. Geologische und biologische Beobachtungen.- Abhandlungen der mathematisch-physischen Classe der Königlich Sächsischen Gesellschaft der Wissenschaften 14, 439-505.

WARREN, J. K. (1982): The hydrological significance of Holocene tepees, stromatolites, and boxwork limestones in coastal salinas in South Australia.- J. Sediment. Petrol. 52, 1171-1201.

WESTBROEK, P., VRIND-DE JONG, E. W. DE, WAL, P. VAN DER, BORMAN, A. H. & VRIND, J. P. M. DE (1985): Biopolymer-mediated Ca and Mn accumulation and biomineralization.- Geologie en Mijnbouw 64, 5-15.

WHITE, W. A. (1961): Colloid phenomena in sedimentation of argillaceous rocks.- J. Sediment. Petrol. 31, 560-570.

WILSON, M. J., JONES, D. & RUSSELL, J. D. (1980): Glushinskite, a naturally occurring magnesium oxalate.- Miner. Mag. 43, 837-840.

WUNDERLICH, F. (1984): Bioturbation auf Raten.- Natur und Museum 114, 14-18.